Raising Turkeys

Beginners Guide
to Raising Healthy and Happy Turkeys

By: Irene Mills

Table of Contents

Part One: Introduction

Backyard poultry has taken off in popularity. Chickens, ducks, and turkey hobby farms have seen a surge in popularity—with good reason! Many people in suburban and some urban areas are raising poultry for food, pleasure, and peace of mind. Chickens' reign supreme, with ducks and turkeys right behind them.

Ranches might raise large flocks of turkeys for consumption and profit while backyard hobbyists might raise a few turkeys just for fun. Both are fun and rewarding.

Why Would You Want to Raise Turkeys?

1. Turkeys provide food. *Really. Good. Food.*
   Turkeys provide food in two ways: egg production and meat. A lot of meat! Everyone thinks of turkeys as Thanksgiving birds, but they can be butchered and eaten at any time. Some varieties will lay a few eggs a week and provide meat later. The timing of your meat production depends on you and your available freezer space.

   While turkeys do not lay eggs as often as chickens, they do lay a larger, protein-rich egg that is very tasty. Since they are much rarer, they sell for a bit more than a dozen chicken eggs. A dozen organic turkey eggs can sell for $8–$10, whereas organic free range chicken eggs sell for $4–$6 per dozen.

There is nothing like gathering your own eggs and harvesting your own meat. The taste and quality are far superior to anything you can buy in the grocery store.

2. Turkeys are great for your garden and flower beds.
   Turkey droppings are high in nitrogen, phosphorus, and potassium which bodes well for plants in the garden and flower beds. With care, it could be of huge benefit, but should be done correctly or you run the risk of damaging plants. We will talk about this more in depth in Part Three.

3. Turkeys create less mess.
   Turkeys are less messy than chickens. Turkeys do not feel the need to scratch like chickens do. They are much more regal. Toms tend to strut around, and females will forage. Neither are destructive toward flower bed mulch or the ground.

4. Turkeys help provide pest control.
   Turkeys are not known for this, but they will provide some pest control for your space. The females are more active than the males, and while they eat bugs and other pests, they tend to be slower and less excited. If you're looking for bug control, chickens might be your best bet; or, look at raising Royal Palm turkeys who are known to be great foragers.

5. Turkeys can provide a wonderful experience.
   If you have kids, they will learn about the circle of life and know first-hand where their food comes from.

There is a natural joy for a child when they can take part in caring for another creature. Turkeys tend to be social and gentle, making them great farm companions for children and adults. On our farm, we knew animals weren't going to be pets but food. That said, our turkey, Mr. Belvedere, is a family favorite and will be extremely hard to say goodbye to when Thanksgiving comes.

There is also the bonus of instilling a good work ethic in children. Turkeys are small enough to be manageable, unlike large livestock like cows or horses. Children will learn many lessons and take pride in their homestead. The entire experience can be a way to strengthen the bond of the family.

6. Personality!

Turkeys have wonderful personalities, are normally social, and dare I say beautiful in their own way. Turkeys rarely show aggression. Toms' strut and puff their feathers, making clicking or "hissing" noises, but they aren't known to charge in an aggressive manner. I've had small roosters that weighed far less who were terrifying as they'd attempt to run up your leg the minute you opened the coop. My turkeys have always been happy to see me, greeting me with a gobble and then shadowing me around the yard. They are more like golden retrievers than attack ninja assassins.

Turkeys develop loyalty with their humans. Ours follow us around. If we are outside, Mr. Belvedere is there with us, puffing out his chest and gobbling in agreement with everything we say.

Fun Turkey Facts

1. Turkeys were the center of Mayan celebrations and offerings. They used them for feasting and even made turkey tamales!
2. In the 1800s, the term "drumstick" was used to describe a woman's leg.
3. Turkeys are all star athletes. Wild turkeys can fly short distances, run up to 12 mph, and swim! Most domestic turkeys cannot fly which was used in great comedic effect during a famous Thanksgiving episode of WKRP in Cincinnati.

4. Benjamin Franklin wrote to his daughter, Sarah Bache "For my own part I wish the bald eagle had not been chosen as the representative of our country. He is a bird of bad moral character. He does not get his living honestly…For in truth,

the turkey is in comparison a much more respectable bird, and withal a true original native of America."

5. In 1963, U.S. President John F. Kennedy spared a turkey, a trend followed by several presidents including George H. W. Bush, who officially began the "Presidential Pardon" of the turkey in 1989.

6. Wild turkeys are thought to be very clever and cunning.

Frequently Asked Questions

Turkeys are not as common as chickens. Therefore, people always have questions. Below are a few of the most frequently asked questions regarding the upkeep of turkeys.

What Are Turkeys Called?

Baby turkeys are called "Poults."
Young turkeys are called "Jakes."
Adult female turkeys are called "Hens."
Adult male turkeys are called "Toms."
A group of turkeys is called a "Flock" or a "Rafter."

How Much Time and Work Will It Take to Keep Turkeys?

There is a fair amount of work to do when you set up. You'll be researching, learning, and making a lot of decisions about coop placement, breeds, the number of turkeys to have, and getting the coop and the feed set up.

After the setup, you may have baby turkeys and you may go into poult maintenance mode. Poult maintenance is far different than adult maintenance and we will cover that a little later.

Once the turkeys switch from a brooder to an outdoor coop, the amount of time spent caring for them drops significantly. However, turkeys are social creatures. You should spend time with them every day. Think of them as feathered Golden Retrievers.

Here's a sample schedule for <u>adult</u> turkey maintenance:

- <u>Every Morning</u>: Let your turkeys out of the coop and into their enclosed run. Change their water, feed them, and check for eggs. A solar door is an amazing coop addition. They are very handy when you've had a long night and want to sleep in.

- <u>Afternoon</u>: Check for eggs again (if a hen is laying, she can lay anytime). Offer a treat like meal worms or fruit scraps. If it's hot they will appreciate cool water.

- <u>Late Afternoon/Early Evening Before Dark</u>: Lock them in their roost. Turkeys, like chickens have the instinct to roost when the sun starts going down. They are also creatures of habit. Once they start roosting somewhere it's hard to get them to stop. Turkeys have been known to fly into trees, on to rooftops, on top of cars, and sheds. Sometimes they don't want to be "cooped up" no matter how nice or clean the coop is.

  Mr. Belvedere prefers to roost on the rail of our metal trailer and will only leave his perch when it's snowing or raining. Mrs. Belvedere must be carried into the coop as she will only go in to lay an egg and then come right out. We worried she would get frostbite, so we literally had to carry her in.

- <u>Once a Week</u>: Clean the coop. If you allow your turkeys to range freely on your property or yard, the clean up won't be nearly as bad.

Another coop cleaning method is called the "deep litter method." Deep litter decreases cleaning the coop to every six months. The idea is that instead of removing the turkey manure regularly and composting it separately, you let it compost in the coop and then clean it out and start over. Four to six inches of dry wood shavings will last six months. Instead of cleaning the coop, all you must do is add more bedding along with some treats to encourage the turkeys to peck, scratch, and aerate the compost. The compost from this process is fantastic but requires time to burn off nitrogen. Do not put turkey droppings directly on plants or they will burn your crops. There are pros and cons; we will discuss the details of this later.

Are Turkeys Expensive?

Turkeys cost more than chickens, eat more because of their rapid growth, and require a bit more time in the beginning.

The initial cost of turkey poults (chicks) is roughly $6 for Broad Breasted Whites and $10 or more for a heritage breed. We will help you decide which turkeys are best for your flock a little later.

Food varies by type and quantity but will range between $12–$25 per 40 lb bag depending on if you prefer regular or organic. Turkeys eat about half a pound of food per day. If you have a flock of 10 turkeys, you will need a bag of feed every eight days or roughly four bags per month. Only you can determine if that cost makes sense for your budget.

It's also good to note that with chickens you can supplement their daily food with kitchen scraps and foraging. Turkeys require a higher amount of protein than table scraps provide. With turkeys, scraps are a treat instead of sustenance.

Do Turkeys Smell?

Turkeys do not smell but their poop and enclosure sure do. Keeping their area clean is effective in fighting the dreaded poultry smell. We will discuss the various ways of cleaning a turkey coop or enclosure later.

How Do I Know How Many Turkeys I Can Have?

This is a two-part question. First, how many turkeys **should** you have versus how many turkeys **can** you have?

Let's start with CAN. Depending on where you live will determine if you can have turkeys. Rural or farm areas most likely have no limitations on poultry or noise ordinances. It's important to know what is allowed in your zoning. When you purchased your home, you should have been given paperwork on zoning or neighborhood covenants for your area, city, or county.

In more populated areas there are neighborhood covenants or Homeowners' Associations (HOAs) which govern their own areas. There are city, county, or parish rules as well. For example, my address in Colorado Springs, Colorado, is legally in the county versus the city. Turkeys are not allowed in the city but "a reasonable amount of domesticated poultry" is allowed in the county. Their reasonable amount is 10 birds with no roosters. Roosters are generally noisy and create noise complaints. Common sense tells me that if roosters aren't allowed, turkeys probably wouldn't go over so well. Male and female turkeys produce sound. A lot of sound!

The hot topic of HOAs is if they allow turkeys or not. The only HOAs I've ever seen allow turkeys are the rural planned communities with 5+ acre lots. Our turkeys reside in Albuquerque with my brother on two acres in the country where they can be as loud as they want.

Bottom line: even if your neighborhood allows poultry, that doesn't mean you should run out and buy turkeys. Think it through. There's nothing more upsetting than having to get rid of your birds because they aren't welcome in your neighborhood.

SHOULD you have turkeys? Do you have enough space for turkeys? Everything is larger with a turkey. They need more space to roam, a larger coop or enclosure, and eat more. With chickens you can get by with a standard 4x4-foot coop from Tractor Supply or Amazon. This is not the case with turkeys. They are six to ten times larger! Enough space for the turkeys and your happiness is essential. A small space needs to be cleaned more often and can create unpleasant odors. Turkeys require movement to build muscle. Four standard turkeys should be in a run no smaller than 16x16 feet. The standard is three to four square feet per bird but the more space you can offer them the better.

Can I Have One Turkey?

No. Turkeys are social creatures by nature and require companions. You should have at least two.

How Many Eggs Will I Get?

Depending on the breed of turkey you choose, you will receive no more than two eggs a week. Most breeds will lay during certain times of the year. If you are looking for egg production, you may want to supplement with chickens.

Can People Get Sick from Backyard Turkeys?

Without a doubt. Yes. You can get sick from turkeys or any kind of backyard or farm animal, but there are MANY safeguards and strategies to avoid getting sick. Cleanliness is key!

The three major points are:
- Wash your hands,
- Use outside-only shoes, and
- Don't let your turkeys inside.

Wash your hands thoroughly with soap and water after collecting eggs or handling the turkeys. You can hold them, have them on your lap and touch them, but you must wash your hands afterward. Turkeys are much larger and heavier than chickens so children should be supervised.

Use hand sanitizer if soap and water are not available. Make sure that children do that as well and supervise them to ensure they get the job done right.

Do not allow your turkeys or baby chicks to come into the house. (They walk around in their own poop and have bacteria on their feet and feathers. They may also peck something in your house and leave bacteria on it.) Turkeys are very curious by nature. It's not uncommon to find them at your back door looking through the glass to see what's going on in there.

Have a pair of outdoor only shoes for each member of the family by the back door. You don't want to track in the bacteria. Teach children that they absolutely cannot go in the turkey enclosure barefoot. Turkey poop can contain worms which can transmit to human feet. Gross!

You can hold your turkeys, but don't cuddle them next to your face or kiss them. (This is more of a problem with chickens than turkeys but err on the side of caution.) I've found that children are very intimidated by the size of turkeys. Especially the toms with their funny faces.

Both turkeys and chickens peck around their droppings. Their beaks, feet, and feathers can have bacterial contamination.

Clean the feeders, water containers, and any other equipment outdoors. DO NOT do this in your kitchen sink.

Collect your eggs once a day. Do not leave your eggs in the egg box or on the ground more than a night.

If your eggs have dirt or debris, you can use a designated piece of fine sandpaper or a brush to get it off. The open pores of the egg will *pull the bacteria in* from the outside. *The CDC notes that you should not wash the eggs because "colder water can pull bacteria into the egg."* We will discuss more details in the egg section later.

Of course, cook your eggs thoroughly.

Part Two: Getting Started

Anatomy of a Turkey

The turkey is a large bird in the genus Maleagris, native to North America. They have been clocked to run up to 25 mph and some varieties can indeed fly.

Turkey Anatomy 101

Toms, or male turkeys, when allowed to grow to full maturity, are a thing of beauty. They puff and strut like peacocks, and in our opinion are just as beautiful. Their parts may look strange and intimidating but all serve a role.

Females, or hens, are smaller in size and do not possess show feathers like the toms. They are more refined.

Snood

What is that thing? Why does my turkey have that long dangly bit? Is my turkey normal? These are all questions I've heard about the snood. The snood is a fleshy appendage that hangs over the beak. It looks like a miniature trunk but just hangs there attracting mates. Seriously, that's its purpose. All turkeys have them, but the male's snood grows with maturity. The longer the snood, the more the ladies are supposed to be attracted to him. They can grow up to five inches when trying to impress a female, and the toms with the brightest snoods tend to attract the most girlfriends. Studies also show that the turkeys with the more robust snoods have better health.

Snood fully exposed.

Snood at rest.

Wattle

The wattle is a flap of skin hanging under the chin connecting the throat and head. Both toms and hens have wattles, but they are definitely more pronounced in the males. Another sign of virility in males, it's used to attract mates and warn away potential rivals. Chickens have two wattles hanging down while turkeys only have one.

Turkey wattle and caruncle.

Caruncle

The warty protuberances on the head of a turkey are called caruncles. On turkeys, the term usually refers to all the flesh that is not a wattle or a snood. Behind each eye, you will see a pea-sized caruncle which are the turkey's ears. The bulbous hunks of tissue behind the wattle are called the major caruncles. These look like brains or intestines. The larger the caruncles, the more testosterone a tom has. When a turkey is feeling frisky, the caruncles around the head and neck turn bright red, while those on the face turn a brilliant aquamarine blue.

Female turkeys also have caruncles, but all three are smaller and remain a less noticeable fleshy pink color.

Beard

As male turkeys mature, they developed a clump of slender, fibrous feathers in the center of their breast, which is referred to as a beard or a tassel. A turkey's beard resembles a horse's tail, except it's shorter and on the front of the body. Like the fleshy head appendages, turkey beards are believed to be an aesthetic feature intended to attract mates; beards become more erect when a tom is aroused. The longer the beard, the older the turkey—they can eventually grow to about 12 inches in length. Rarely, toms have multiple beards, which are aligned vertically along the breast, with the longest beards on the bottom and getting smaller as they go up. Some, but not all, female turkeys also grow beards, though they top out at about six or seven inches.[1]

Feathers

Feathers on both male and female turkeys are essentially the same. They have short, downy feathers on their breasts and back for warmth. They have several layers of wing feathers, which provide loft for flight for some breeds (more on that later): the outer layer is made up of "covert" feathers; then comes a layer of slightly longer "secondary" feathers; the longer "primary" feathers are closest to the body and hang down almost to the ground. There are two layers of tail feathers: the short "coverts" in front, and the much longer "rectrices" behind (up to 18 inches long), of which there are typically 18.

Female turkey feathers are usually white, black, or gray, perhaps with a few mottled markings here and there. On males, however, markings are much more pronounced and colorful, and include the characteristic band patterns on the tail feathers.

A Tom in all his feathered glory.

Males weigh between one-and-a-half to two times as much as females, but they appear even larger because of their ability to puff out every feather on their body—which is known as strutting. Strutting is common behavior throughout the year but is almost constant during the spring mating season. When strutting, the males flap their wings, fan out their tail feathers in full display, and make a "gobbling" noise—making it very easy to

---

[1] ModernFarmer.com

tell who is pursuing who.[2] Our turkey, Mr. Belvedere, struts constantly and puffs out every chance he gets, to the dogs, the goats, and every human who visits.

Spurs

Male turkeys have spurs on the backs of their feet. Jakes (young teenage turkeys) will have about a half-inch nub. The older the turkey the longer the spur. Adults can have up to two-inch spurs. Spurs are used for fighting. Especially during mating season.

Turkey spur[3]

Decide How Many Turkeys You Will Have

---

[2] ModernFarmer.com
[3] Thebirdwatchersstore.com

It is easy to calculate how many turkeys you can have.

Turkeys require at least 25 square feet of space to roam in an enclosure. That's not a lot of room and for their happiness, bigger is better. For four turkeys I would recommend at least a 25x25-foot enclosure. Do you have that much room? If not, maybe your turkey plan should be put on hold.

First, you will need to measure the total space you have for a coop and enclosure and calculate the square feet.

Here is the most straightforward estimate:

If you want a flock of six turkeys, you need a coop with a minimum of 24 square feet in total. You will need a 6X4-foot coop with a door large enough for a full-grown tom and at least 400 square feet in the enclosure *outside* the coop. It sounds like a lot but it's only 20x20 feet.

Why do turkeys need so much space? They're large! They also get bored. When boredom strikes, they begin to pick on one another which can materialize in pulled feathers, injuries, or cannibalism. Yikes!

Happy turkeys in their pen with a simple homemade roost.

Again, the more space you can dedicate to your turkey flock the better.

If you are a beginner, we recommend that you start small. If you have enough space for 12 turkeys, that's great, but still start small. Start with half the number you can fit in the pasture and coop and see what you think of the process before jumping in.

Ask yourself, how many turkeys *should* I have versus how many turkeys *can* I have? Decide on a number and stick to your plan. You won't be the first person to see how cute baby turkeys are and decide you need a few more.

Contact a Local Expert if Possible

Guides such as this one, as well as books, articles, websites, and forums, are very important and useful to enable you to get solid knowledge for raising your turkeys. There is a lot of reliable information available. People can successfully raise turkeys by doing their research, buying, or making a roost and enclosure, and purchasing the chicks they want.

You can do this! Any knowledge you need is merely a search away.

Even so, nothing can replace the information gained from speaking with and watching a local farmer or current keeper. Being able to ask questions is priceless. One of the best forums I've found is on Facebook. Search for a farming or poultry group and join! The information exchanged is amazing and offers a quick invaluable source. Talking to others within your community forums is also a great source for finding turkey poults for purchase.

If you are going to purchase from a store, ask questions, and find out where the chicks have been bred and raised. You want to ensure that they are from a trustworthy source so that you are not getting diseased chicks.

Basic Requirements for Turkey Chicks (Poults)

No matter what kind of turkeys you choose, you need these necessary things:

- Brooder
- Heat Source
- Thermometer
- High Protein Food
- Water
- Bedding

An enclosure is also called a brooder. Think of it as a nursery for the poults.

The brooder setup for raising poults is like one for baby chicks (chickens). The biggest difference is that turkeys require a higher temperature than chickens. If you attempt to raise turkey poults and chicken chicks at the same time, it's recommended to raise them in separate enclosures.

Baby chicks in their comfy brooder made out of a galvanized stock tank.

An enclosure can be anything with a solid, draft-free wall. Turkeys will grow and begin to jump around so we recommend a container that offers them space to grow without being able to jump out. It's also recommended to put a wire or mesh screen or lid over the top. This keeps adventurous poults in and curious creatures (cats, dogs, or small children) out. Stock tanks, tall wading pools, or large boxes work well.

Brooder set up with heat, food, water, and cover.

Once you have their home established it is time to make their stay comfortable. Heat, food, and water are essential for their survival.

Heat

A heating pad or lamp is needed to maintain a temperature of 95–100 degrees. Turkey poults cannot maintain their own temperature, so they need a little help. Maintain the 95-to-100-degree temps for 10 days. Then drop the temperature five degrees every week until the poults are eight weeks old.

Unless you live in a very cold climate, your poults should be able to live outside on their own at this point. Most of the time poults are hatched and sold in the spring. If it's still 45 degrees or less outside during the day, you might want to keep your poults inside a week or two longer.

Food

Young turkey poults require a crumble containing 30% protein. They grow rapidly and need all that protein to grow properly. If you want a nice juicy turkey for Thanksgiving dinner, you must feed it well.

Poults need to eat 30% protein crumble until they are eight weeks old. Then they drop down to 20% protein. When the females begin laying, you may also supplement with special laying feed.

Food can be sourced from farm supply stores, feed stores, and even online. There are many types and varieties. Those shown are from Murdochs.com.

Water

Turkeys require fresh water at all times. This may prove difficult in winter but there are several options to ensure your turkeys have a water source.

Turkeys tend to keep their water a lot cleaner than chickens. They don't scratch around like chickens and debris doesn't get tossed in. Still, there are several types of water containers and setups to choose from. Personal preference prevails.

Pans and buckets – The standbys are shallow pans for poults and jakes. Nothing too deep that they can fall in and drown. Yes, this truly happens. I put a large rock in the center of my buckets for this very reason.

Our favorite bucket is heated. It requires electricity to the fence but keeps the bucket warm enough to not freeze.

Automatic waterers – Large operations will want to install an automatic waterer. They are hooked up directly to the hose and have a float or mechanism that triggers the filling of the bucket when it reaches a certain level. These save a lot of time but, like any other bucket, must be cleaned out every now and then.

Ground automatic waterer.

Hanging water buckets – This is my method of choice in the summer months, and it works well. I use a five-gallon bucket from the hardware store which has been modified with five water "nipples." Water only comes out when a bird touches it. It keeps the water very clean and holds up to five gallons at a time. The only drawback is it can only be used in above-freezing temperatures.

Water bucket with chicken "nipples."

Buckets can be sourced at any hardware store, and we found the chicken nipples on Amazon. They are also located at feed and supply stores. Very easy!

Basic Requirements for Adult Turkeys

- Perches/Roosting Bars

A wooden roosting bar off the ground feels right for them. Their instinct is to sleep on a high branch. Both wood and plastic perches can be bought but turkey feet find it easier to grip wood. Trust me, they don't care if you buy a fancy perch or use a cheap two-by-four from the backyard or home improvement store. Perches should be large enough for a turkey to rest easily without teetering off. My chickens sleep happily on a two-inch wooden dowel that was repurposed from a closet remodel. Our turkey, Mrs. Belvedere, sits on a two-by-four placed so it sits wide. Mr. Belvedere prefers a metal railing on our trailer parked in the yard.

*Warning! You can create the Taj Mahal of turkey palaces for your birds, and they will decide to sleep on top of your boat. Or the doghouse. Or on top of the BBQ grill. My point is, turkeys go where they want to go and it's very difficult to change their minds. If you do not want them to choose a roost on their own, you will need a fence to contain them within a certain area.

How much room does each bird need?

Each turkey needs 18"–30" of space on a roost. Also, note that if you are housing chickens and turkeys together, the roosting place for turkeys should be the highest point in the coop. If you have a non-flying breed, you may want to offer some sort of ramp for them to climb up.

Choosing or Building an Enclosure

This custom turkey coop was built with plans found online. Note the hardwire and enlarged door!

Turkeys are much different from chickens when it comes to their housing and pasture fencing needs. Adult turkeys prefer to be outdoors. They are hardy and tolerant of many different weather conditions, so they can be kept outdoors most of the time beginning from the age of eight weeks.

Coops

Once you have the measurements for how large your coop and enclosure will be, you need to decide whether to purchase a coop, build yours from purchased materials, or put together a coop with recycled materials.

Poultry coop. Note the various birds with enough room for a big tom!

Whichever method you decide; the enclosure must be larger than a standard chicken coop. Do not buy a chicken coop for your turkeys unless it's large enough to accommodate them. These birds get big!

Coop made from PVC and wood.

Normal chicken doors are not large enough and will tear out feathers or your turkey will choose to remain outside and not even use it. Either way, it's not fair to your birds.

Building Your Turkey Enclosure

You can build your coop with a plan, repurpose/recycle something else to become your chicken coop, or build your coop with recycled materials.

Whatever plan you choose to build your coop, DIY can be an option to keep costs down. It doesn't have to be hard. It can also be an opportunity for a family project leading up to getting your turkeys.

In this section, we will look at:
- Building your coop from a plan.
- New and used materials for building your coop.
- Ideas for repurposing or recycling something else to be your turkey coop.

Building Your Coop from a Plan

Whether or not you have building skills or experience, numerous helpful sites offer free turkey coop plans. They are so numerous that our job in this guide is not to find them but to curate out of the overwhelming number of resources.

Here are some of the websites we found simply by searching "DIY turkey house". As you will see, the search defaults to chicken coops. You must look at sizes and figure out what works for your flock. With a large enough coop, it's merely a matter of creating a larger door and nesting box.

Etsy – Etsy overwhelmingly won the search for coop plans. There are so many types, sizes, and prices to choose from.
eBay – Many plans to choose from here as well.
The Garden Coop – Very pretty selection to choose from. This seemed to cater to those with building know-how.
Bonanza—Just Build It – One of the only sites that offered a printed version as well as a download.

Materials for Building

The primary criteria for coops are:
- Are the materials sturdy and durable?
- Are they strong enough to provide protection from predators?

- Are they non-toxic?
- Will the material attract pests such as mites?
- What is the cost?

Wood is the most common material used for coops. Hardwoods such as tropical wood, redwood, or cedar (we will talk more about cedar) are more pest-resistant than softwoods. Most coops end up getting built with the more affordable softwoods such as pine or fir and then stained or painted with a non-toxic product. A disadvantage to wood is that it's a porous material and retains moisture, bacteria, and smell.

Coop made from wood.

Pressure Treated Wood

If you use recycled wood, you don't want chemicals going into your eggs and chicken meat. Wood that is older than 2003 has chromate copper arsenate (CCA), which includes arsenic. CCA is very poisonous, and "long term exposure to the arsenic, which is found in some types of CCA-pressure-treated lumber, can increase the risk of lung, bladder, and skin cancer over a person's lifetime."

It's not only long-term exposure that can be a health hazard. "The arsenic levels in the wood have also been shown to have a negative health effect…for example, when children touch wood in play areas, and then put their hands into their mouths." Coops made to spec by a plan with new wood are very attractive, but you can also use recycled wood to reduce the cost. Just heed to the warnings above on the wood you are getting.

Pallets are wonderful and most often free sources of recycled wood. Old barns and buildings that have been torn down are also great. Where do you find such treasures? Online marketplaces are a great source.

Habitat for Humanity operates ReStores where you can find just about any building supply for extremely reasonable prices as well. You will find lumber, hardware, electrical, doors, windows, and just about anything else you can think of that you might need. All supplies are donated, and the revenue funds the Habitat for Humanity projects. Definitely, one fun resource!

Repurposed super cool items that have been used in the past are:
- Metal shipping containers
- Old cars (stripped and emptied of hazardous liquids)
- Buses
- Water tanks
- Grain silos
- Trampoline frames
- Boats

Coop made from an old shipping container.

The thing to remember if you are repurposing is that you need to follow basic coop protocol. Remember…

- You want the roost to have a perch and be the highest point.
- The nesting boxes need to be dark and have built-in access to collect eggs and clean out the bedding.
- Make sure that the repurposed item allows for enough space for the birds you want or that you limit the number of birds to the size of your super cool repurposed item you found. Either way, don't crowd your flock.
- Add a layer of insulation to ensure it's warm enough for your area. Colorado temps can get into the sub-zero temperatures. We make sure our flock has a foot of pine shavings and insulate them from drafts and wind. The birds are nice and toasty!

Feeders

Feeder design is a personal choice. I like an indoor feeder during the winter as it avoids rain and snow accumulating in the food bowls which then makes the scratch soggy. In the summer, I use a homemade trough system. It's raised up off the ground and doesn't allow the birds to scratch it out. It also evens out the distribution of morning scraps which the turkeys go crazy over. It allows more turkeys and chickens to eat at the same time and less birds get "henpecked" or chased away. I have absolutely no data on this, but I swear it creates more harmony within the flock.

Feeders come in all shapes and sizes.

Figure 1 Turkey Feeder, Solwayfeeders.com

Poults will need feeders closer to the ground. As they grow and eat more, you will want to switch to a larger feeder.

There's nothing wrong with old pans, dog food bowls, or shallow buckets if you don't mind feeding the turkeys every day.

If you want something larger that holds a bag of food or more at a time, there are options for that as well.

Trash cans or storage boxes with holes cut out work well. As long as there is a base of some kind to catch the falling food, you will find that style is very efficient.

There are tons of plans and photos on Pinterest and Google showing how to build a PVC gravity fed feeder. It holds as much as you want and avoids feeding the birds daily. Side note! You should always check your food source daily. If a critter gets in, it can decimate your food supply without you knowing. Turkeys rely on that high protein calorie count to grow large and delicious. You'll want to keep the food coming!

Figure 2 PVC Feeder, Instructables.com

Water

Important advice about waterers: You must make sure that you have several sources of water. As with the feeders, the less dominant hens need access without restrictions, and in the case of any automatic system, there may be a mechanical problem causing malfunction. Both availability and back-up are essential for your turkey's access to water.

Are you in a cold climate? There is nothing worse than having to fill a water bucket multiple times in freezing conditions. Anyone who has ever tried to get a block of ice out of a bucket knows what I am talking about.

Water buckets with built in heaters are the way to go. There are two types: electric and solar. I've always used electric as I'm fortunate to have electricity at the coop. Besides making sure it's full, I've never had issues with it freezing over. On very cold and nasty days where the flock refuses to leave their enclosure, we leave their food and water inside the covered run where we've also stapled contractor grade plastic to the outside of the coop. There's enough air circulating to ward off frost bite but enough coverage to keep the flock above twenty degrees.

It's important to note that many feathered birds live outside with no heat or comforts from a coop. They are built to withstand the severe cold. To a point. I don't believe in allowing my animals to suffer. When the temps dip below ten degrees, I step in and help raise the temperature to avoid frostbite on their feet and faces.

Toys

Birds love toys. Especially if they involve food. Google "chicken toys" and you will find all sorts of fun contraptions that can cross over for your turkeys. Why are toys important? Toys create stimulation for your birds.

Flocks who are "cooped up" become stressed out and resort to all sorts of nasty behaviors like feather pulling, bullying, or even cannibalism. Let's face it, it's also fun to watch your flock enjoy the toys or stimulation activities. Here are a few that we bought or created for our chickens and have witnessed the turkeys using as well.

Hanging cabbage – This is a favorite activity with any veggie.

Cricket tubes – Purchase online or make your own; it's exactly how it sounds. Drill some holes in a PVC pipe or hollow log, seal one end permanently and fill with crickets before closing the other. Waiting for crickets to come out will entertain them for hours!

Newspaper feast – Sprinkle meal worms or berries in between the layers of newspaper and watch them search and scratch.

Pumpkins or watermelons – Smash the fruit in half and watch them go crazy!

Lawn clippings/leaves – They love freshly cut grass and piles of leaves.

Dust Bath

Dust baths are a necessary and natural process your turkeys will happily partake in. If you have natural dirt in your run, they will take a dust bath by instinct. Warm days trigger the instinct, and you will see your birds scratching down into a hole where they scratch and flail dirt all over their puffed-out feathers. They will do this for quite a while and then shake off—nice and clean.

This is their natural way of combatting oil and dust mites. It's fun to watch. Especially when a group of them do it together. The dirt flies!

If your flock does not have natural dirt, you will need to provide "dirt" for them. This can be done in an old tire, wading pool, or a wooden box sunk into the ground. Think of it as a sand box for your flock.

Fill it with playground-rated sand, dirt from another area of the yard, or bags of chemical-free soil from the hardware store.

Some people add dematiaceous earth to it to provide another healthy layer to their bath but it's not necessary unless you suspect a mite infestation. We'll discuss dematiaceous earth and mites a little later when we talk about health issues.

Bedding

What are the functions of bedding in your coop?

- In the roost
    - Catching the poop (turkeys poop all the time, including a lot at night).
    - Bedding provides insulation.
- In the nest boxes
    - Bedding is there to be comfy, warm, soft, and the best feeling for the hens to want to lay eggs in a convenient location.
- In the covered area of the coop
    - Bedding catches and absorbs the poop and makes it an excellent place to scratch and hang out if your turkeys need to be inside for protection or don't want to be outside.

Generally speaking, the main criteria for bedding is that it is:
- non-toxic
- absorbent
- quick drying
- compostable
- and relatively inexpensive.

After those criteria, you need to decide what material will be the best for the happiness and comfort of your birds and you. The absorbency and compostable nature of the bedding will make it more comfortable when you clean the coop.

Pine or Other Wood Shavings

You can get wood shavings from a local woodworker or purchase shavings from a feed store. It is absorbent and makes the cleaning a bit easier.

If the bedding stays dry, it will not smell, but if it gets wet (from an overdose of runny poop or spillage from the waterer), then it will certainly smell. Generally, the absorbency of wood shavings means that they don't smell. It is one of the strong points of this bedding.

You may find out that you have a choice of "cut" size for the wood shavings. You want a larger cut for wood shavings as the smaller ones can be too fine and lean toward being sawdust. Filling your coop with sawdust means that you can risk wood dust, which can cause respiratory problems in turkeys.

If you decide to do the "deep litter" method of composting your turkey manure, wood shavings are the best choice as they have enough absorbency. Spread shavings and allow the turkeys to do their thing until they are "soiled" and compacted down. Then dump more shavings on top. Repeat as needed throughout the winter. It creates a natural insulation and begins to compost itself as the nitrogen begins to break down the wood. In the spring, shovel it all out into the compost pile and begin again.

This is our chosen method, and it works very well. The only negative is the smell. It can be bad in the spring.

Tip: If you can source a local woodworker or other business for your wood shavings, then you are more likely to get fresh-cut wood. This is much preferable (especially if you are composting with deep litter) because the wood still has sugars that are active in it. This process helps break down the elements in the poop and composts it quickly.

Hay or Straw

Hay still has nutrition/seeds such as alfalfa that is given to cows. It's usually more expensive than straw and isn't recommended.

Straw is usually easy to find, and it is free or cheap. This attribute makes it an excellent choice for a daily or weekly coop cleaning routine. It's soft. It is also warm and gives a lot of insulation, which can be a big plus if you live in a cold climate.

The main downside of straw is that it is not very absorbent. Straw is not a good choice for a deep litter method of manure compost and coop care; it needs to be replaced often. If it is not replaced frequently, the coop will end up smelling, and that ammonia odor is not just unpleasant but harmful to your turkey's health. Straw gets heavy and clumpy with poop and is not as easy to clean in any case.

If you choose straw, make sure that you source it from someone who can guarantee that there are no pesticides in it. The chemicals in pesticides could kill your turkeys.

Sand

Not recommended in the coop but very nice in the run.

Inside the coop, sand is a no-no. Imagine a cat box. The waste clumps and begins to smell due to the ammonia. Sand offers nothing to break down, so it lingers. Yuck!

In the run it offers a soft landscape for your birds' feet. It's easy to rake or clean and absorbs water easily.

If you use sand, make sure it is **not** "play sand" or sandbox sand. Both are too fine and can cause dust and respiratory problems. They also sometimes contain salt, which is not healthy for your turkeys except in measured doses.

A pile or large pool of sand in the run is an excellent choice for grit and baths.

There are some ready-made chicken coop bedding options available. In our opinion they are expensive to use in a turkey enclosure. Due to the size of the enclosures and amount needed they tend to be very pricey.

- **Koop-Clean** by Lucerne Farms. It is a blend of chopped hay and straw with a mineral-based odor-neutralizing ingredient called "Sweet PDZ."

- **Chopped straw**

- **Aspen shaving nesting liners**

- **Premium pine shavings**

Please stay away from the following two materials for your coop bedding:

Cedar Shavings—BEWARE!
There is a chemical in natural cedar oil that is toxic and can give turkeys respiratory problems. Cedar is tempting because you may hear that it does kill parasites. That's awesome, but the downside is that it may kill or is likely to sicken your birds, too.

Sawdust
Sawdust is excellent for compost but not for bedding. It cakes and quickly absorbs too much moisture that doesn't dry out, leaving a stinky, matted mess that is difficult to clean and will soon smell of ammonia and be a hazard to your birds.

In the Outdoor Run/Enclosure

Diatomaceous Earth (DE)

Many poultry raisers consider diatomaceous earth to be a kind of secret ninja move for their flocks.

Diatomaceous earth is found at the bottom of ancient sea beds. It is the skeletons of the cell walls and shells of single-cell marine phytoplankton (diatoms) that have fossilized and, over time, have turned into powder. The DE is ground further into a very fine powder. There is commercial/industrial DE that is sold for many uses, including as a home cleaning agent.

The DE used on farms and specifically in coops is a food-grade DE. This is an important distinction. DE is used as an insecticide for fleas and other insects, a de-worming treatment, and source of minerals.

Coop Bedding Conclusion

There are many types of bedding you can use in the turkey coop. It all comes down to cost and personal preference. The size of the coop and where you live are also important factors. If you live in an area with a colder climate, you need a deeper layer of bedding to add warmth and insulation. Coops in the southern, more humid regions need less due to mold concerns.

The runs also require different maintenance and should be treated differently than the coop itself. Most turkey coops will be set up just like chicken coops—only larger. The runs are also identical (but larger). Turkeys will not scratch as much but they do enjoy many of the same comforts as chickens do.

Choosing Your Turkey Breeds

Choosing your turkeys is very fun, but requires thought before buying whatever is at the feed store or ordering the birds with the pretty feathers. You need to think about why you are raising turkeys, how much space you have, and what kind of climate you live in.

Are you going to eat your birds? Are you raising a barnyard friend? Do you want to breed your own birds? Do you need hot climate birds, or do you live in sub-zero temperatures? Do you have trees for flying turkeys to roost in? Do you have a great expanse for a foraging flock?

These are all great questions to ask yourself BEFORE you make a purchase.

Turkey breeds can be categorized into three categories:

Broad Breasted Breeds
Heritage Breeds
Rare Breeds

Broad breasted breeds are breeds who have been designed by breeders to fill a specific purpose—usually meat consumption. They are raised for one purpose and that is to be eaten. Commercial turkey producers raise broad breasted turkeys because they grow faster and produce large birds. The bigger the bird, the more money to be made.

Butterball is the largest producer of frozen turkey in the world.

Broad breasted turkeys cannot breed on their own and require artificial insemination. Their large size prohibits breeding, and the weight of the hens sometimes damages their eggs. If you're looking for a breeding pair, broad breasted turkeys are not a good option.

Heritage breeds are breeds of turkeys that have carried specific genetic traits, dispositions, and coloring from ancient lines. They have been bred over time to adapt to conditions or geographic areas. Heritage breeds are disease resistant, hearty, and in some cases can fly!

Heritage breeds can breed on their own. If you have a tom and a hen, there is a good chance of her laying and raising poults.

Out of the 250 million turkeys produced in the United States about 1% of those are heritage breeds. Heritage breeds take longer to reach maturity—much longer than the hybrid turkey breeds. Due to the slow growing process, many people believe heritage turkeys to be far tastier than their fast-growing hybrid cousins.

According to the Livestock Conservancy, turkeys must meet all the following criteria to qualify as a heritage turkey:

1. **Naturally mating**: The heritage turkey must be reproduced and genetically maintained through natural mating, with expected fertility rates of 70–80%. This means that turkeys marketed as "heritage" must be the result of naturally mating pairs of both grandparent and parent stock.

2. **Long productive outdoor lifespan**: The heritage turkey must have a long productive lifespan. Breeding hens are commonly productive for 5–7 years and breeding toms for 3–5 years. The heritage turkey must also have a genetic ability to withstand the environmental rigors of outdoor production systems.

3. **Slow growth rate**: The heritage turkey must have a slow to moderate rate of growth. Today's heritage turkeys reach a marketable weight in about 28 weeks, giving the birds time to develop a strong skeletal structure and healthy organs prior to building muscle mass. This growth rate is identical to that of the commercial varieties of the first half of the 20th century.

Rare breeds are breeds of turkey that are threatened, endangered, or almost extinct. They can be heritage or broad breasted but have very low numbers. Many of the breeds are based outside of the United States. They are special breeds who will die out unless man steps in and helps the breed along. If you're planning a backyard flock of turkeys, it's encouraged to source a rare breed if you can. They are beautiful and you're doing your part to keep a beautiful piece of history alive.

Since we are discussing farm or backyard turkeys, we will be discussing domestic turkeys versus their wild cousins.

We will show you the average size hens and toms and rank their egg production, growth rate, and taste of meat on a scale of 1–10 with 10 being the perfect bird.

Heritage Turkey Breeds

Auburn or "Light Brown" Turkeys

Auburn turkeys are stunning birds containing reddish brown feathers with either white accents or a full dark-brown body. They are also one of the rarest heritage breeds in existence. When you think of a turkey on Thanksgiving, chances are you're picturing an

Auburn turkey. Its heritage dates as far back as the 18th century where records of the Auburn turkey were found listed in the receipts for turkeys being taken to the "turkey trot" markets.

The Auburn turkey is in trouble as very few flocks exist. Their close cousin, the Silver Auburn, is easier to find but if you can find a true Auburn or Light Brown turkey, it's highly encouraged to raise and breed them.

Black Turkeys

Norfolk Black, Spanish, or Black turkeys are all the same. They are all breeds of domestic heritage turkeys introduced to European explorers by Spanish Conquistadors. The noble Black turkey was taken home to England where they were bred in Europe for hundreds of years. It is thought to be the oldest breed of turkey in the U.K.

A 1998 census conducted by the American Livestock Breeds Conservancy found that there were only 200 Black Spanish turkeys remaining in the United States, which were being raised by just fifteen different breeders.

There are now conservation efforts in place to revive this amazing breed of turkey. If you're looking at breeding turkeys for fun and not consumption, this is your bird!

Blue Slate Turkeys

The Slate, Blue, or Lavender turkey variety is named for its color, which is solid to ashy-blue over the entire body, with or without a few black flecks. Hens are lighter in hue than the toms. The head, throat, and wattles are red to bluish white. The beak is horn in color, the eyes are brown, and the beard is black. The shanks and toes are pink.

Blue turkeys are hard to find but are making a comeback.

Bourbon Red Turkeys

The Bourbon Red turkey was recognized by the American Poultry Association in 1909. It was ambitiously promoted for its large breast and richly flavored meat. Growers couldn't keep up with the popularity of the broad breasted varieties and interest fell off. Renewed interest in the health and superior flavor of the Bourbon Red has captured consumer interest and created a new following.

Bourbon Red turkeys are handsome. They have brownish to dark red plumage with white flight and tail feathers. Tail feathers have soft red bars crossing them near the end. Body feathers on the toms may be edged in black. Neck and breast feathers are chestnut mahogany, and the undercoat feathers are light buff to almost white. The Bourbon Red's throat wattle is red, changeable to bluish white, the beard is black, and shanks and toes are pink.

The Bourbon Red turkey is an outstanding farmyard bird and meat supplier. With superior taste and light pinfeathers, it makes an excellent choice for small hobby farms looking to produce "pretty" birds for consumer consumption.

Midget Whites are a cross between the Broad Breasted White and Royal Palm. Smaller in stature, they take up less space and can make great farm animals as long as you don't expect them to breed on their own.

Midget White turkey feathers are white. They have pink legs, black beards, and red wattles. Baby Broad Breasted White turkey poults are yellow.

Narragansett Turkeys

Named for Narragansett Bay in Rhode Island, where the variety was developed, the Narragansett turkey is an amazing breed. This turkey is the total package of looks, disposition, and taste. The Narragansett became the foundation of the turkey industry in New England and would quickly grow in popularity throughout the United States.

Interest in the Narragansett began to decline in the early 1900s as popularity of the Standard Bronze grew. The Narragansett was not used for commercial production for decades until the early 21st century, when renewed interest in their health, sustainability, and superior flavor captured consumer interest and created a growing market niche.

The Narragansett color pattern contains black, gray, tan, and white. Its pattern is very similar to the Bronze turkey, with steel gray or dull black replacing the coppery bronze. The Narragansett's beak is horn colored, its head is red to bluish white, and its beard is black. The shanks and feet are salmon colored.

Royal Palm turkeys are the showgirls of the turkey world. Royal Palms were first noted in the 1920s after a farm in Florida mixed the Narragansett, Black Bronze, and wild turkeys together. Mating occurred and their unique coloration was cited.

With their bright white feathers tipped in black and bronze, they are truly stunning.

The saddle is black which provides a sharp contrast against the white base color of body plumage. The tail is pure white, with each feather having a band of black and an edge of white. The coverts are white with a band of black, and the wings are white with a narrow edge of black across each feather. The breast is white with the exposed portion of each feather ending in a band of black to form a contrast of black and white similar to the scales of a fish. The turkeys have red to bluish-white heads, a light horn beak, light brown eyes, red to bluish-white throat and wattles, and deep pink shanks and toes. The beard is black.

Royal Palm turkeys are voracious foragers and are above average flyers. They aren't the largest turkeys on the block, but they make great farm turkeys if you're looking for pest control and personality.

Earlier in the chapter we discussed the Broad Breasted Bronze which is the Standard Bronze distant relative. It's confusing, we know. Basically, it comes down to this…

Broad Breasted Bronze turkeys cannot mate on their own. Standard Bronze Turkeys can mate on their own. Historically, the Standard Bronze turkeys are the most popular turkeys of all time. They are known for their quality meat, tame disposition, and size.

Standard Bronze turkeys were put on the Livestock Conservancy "watch list" as there are only a select number of breeders who maintain small flocks.

With backyard and small farms booming within the last few years, the demand for true Bronze turkeys has grown. Their biological fitness, superior taste, and beautiful black and bronze coloring have made them all the rage once again. The Standard Bronze variety is stately and imposing in appearance. The Standard Bronze turkey has a copper or bronze-colored metallic sheen to its body and a white fringe to its tail.

If you're looking for a healthy flock of turkeys who will mate naturally, require very little upkeep, and offer a delicious bird for consumption, the Bronze turkey is for you.

White Holland Turkeys

White Holland turkeys are the rarest and most difficult turkey to authenticate. Once the most popular white turkey in America, its popularity waned when producers switched to its faster and more robust cousin, the Broad Breasted White turkey. Today, the Broad Breasted White remains the number one seller and the White Holland is all but extinct. Its numbers are extremely low, and a true White Holland turkey is nearly impossible to find. White Holland stock has been mixed with the Broad Breasted White. Its cousin, the Britain White, is also in danger of being "bred out" as more and more breeders hope to capture the fast-growing capabilities and size of the larger more robust Broad Breasted White while keeping the flavor and biological superiority of the Holland White.

The White Holland turkey is showy in appearance, with snow white feathers and a red to bluish head. The beard is black, the beak is pink to horn colored, and the throat and wattles are pinkish white. Shanks and toes are pinkish white, and eyes must be brown. For historical purposes the Holland White should be saved. It maintains an important piece of agricultural history and can be a wonderful addition to farms and homesteaders.

Broad Breasted Turkey Breeds

Beltsville Small White Turkey

The Beltsville Small White turkey was created by the USDA's Beltsville Maryland Agricultural Center in 1934. It was bred to create a small white turkey with no dark pinfeathers. The Beltsville Small White turkey is known to be gentle and docile. They are good egg producers and good mothers but cannot breed on their own.

The plumage is white, with the head red to bluish white. The beard is black, the beak is horn colored, and the eyes are dark brown. Shanks and toes are pinkish white.

The Broad Breasted Bronze turkey was created by breeding European turkeys with wild turkeys in North America. There is much confusion between the Broad Breasted Bronze and Heritage Bronze. Due to size, neither can produce on their own, and while the Standard Bronze has retained its heritage classification, the Broad Breasted Bronze has not. The Broad Breasted Bronze dominated the commercial turkey industry for years but was replaced with the Broad Breasted White due to it being a cleaner looking bird.

Both Bronze turkeys are listed on conservation lists as "priority." The Broad Breasted turkey is a beautiful breed. Very regal and huge! If you're looking for a meat bird with the classic barnyard turkey, this is it.

Black feathers with a large band surround most of its body. Wide yellow feet, a red wattle, and blue and red head make up the coloring of this bold turkey.

*Cannot reproduce themselves. Hens will lay but due to the enormous size and weight of the toms, they can't fertilize the eggs. Eggs should be picked up as soon as possible as hens tend to break them.

Broad Breasted White Turkey

Broad Breasted White turkeys were created by breeding White Holland and the Broad Breasted Bronze turkeys. It has become the number one turkey on the market and is prominently displayed on the table at Thanksgiving.

Broad Breasted Whites are preferred for processing because they produce a cleaner looking carcass. Their feathers are all white which means their pin feathers are not visible when the bird is plucked and dressed.

The Broad Breasted White turkey has been bred with one goal in mind: Size! Some weigh over 50 pounds full grown! Some people feel that taste and texture have been ignored for size and commercial profit. If you want to raise birds that taste and look similar to the turkeys bought in the grocery store, you should definitely grow Broad Breasted White turkeys.

Broad Breasted White turkey feathers are white. They have pink legs, black beards, and red wattles. Baby Broad Breasted White turkey poults are yellow.

*Broad Breasted White turkeys cannot reproduce themselves; they require a little help. Hens will lay but due to the enormous size and weight of the toms, they can't fertilize the eggs. Eggs should be picked up as soon as possible as hens tend to break them.

Rare or Endangered Turkey Breeds

Over the last 100 years, turkey populations have been on a steady decline. With the creation of Broad Breasted varieties, the need for smaller turkeys has slowly waned. As the general population craved larger varieties, the farmers began raising less and less heritage or rare varieties of turkey.

The past 10 years has seen the resurgence of the hobby farm and farm-to-table movement. With it, a renewed interest in rare breeds has kept many of the rarer varieties alive. Sadly, the Turkey Conservation has counted less than 15 flocks of these rare creatures in the world.

Dindon du Gers Turkey

Dindon du Gers or French Black turkey, is a large bird with black feathers and bronze, green, and brown tones on its back. Hens are reliable brooders who will gladly adopt the poults of other hens. Sadly, they are almost extinct.

Dindon rouge des Ardennes Turkey

Dindon rouge des Ardennes turkey is a beautiful buff colored turkey which supposedly was brought to Flanders in the sixteenth century from Mexico by the Spanish.

The breed almost disappeared, but started again in 1985, especially in Champagne-Ardenne. Today it is the most popular turkey breed in France.

Midget White Turkey

In the 1960s, Midget Whites were bred with the goal of being a Broad Breasted White turkey, but smaller. They were developed by the University of Massachusetts.

It was believed there would be a market for this variety, but once the turkeys were available, they didn't get the attention initially anticipated.

As the variety was dying out, a former student named Dr. Bernie Wentworth began refining the variety that we know today.

Midget White turkeys are still in danger of dying out completely, and the only individuals that are working to preserve the variety are private breeders. The Midget White is known for its docile friendly personality, egg production, size, and meat production.

White Holland Turkey

White Holland turkeys were originally an important commercial bird in the U.S. in the early 20th century, but they have since been replaced by faster growing and larger varieties. Today the White Holland turkey is considered a Heritage breed and is very rare—listed as threatened by the Livestock Conservancy—and mostly kept by exhibition breeders.

Part Three: Maintenance

Turkey Feed and Nutrition

Turkeys, like any other living thing, require food and water. They're simple creatures but do require a little tender loving care. The better you treat them, the better tasting your turkey will be.

Water

Turkeys require clean, fresh water every day. Anyone who has ever raised poultry knows keeping the water clean is the hardest part. The good news is turkeys tend to do a bit better as they don't scratch as much.

Water can be as simple as a low bucket on the ground or a more complex set up that delivers a constant flow. Both have good and bad qualities.

Buckets tend to get spilled, require constant filling, and can become heated in the summer. They are also economical and can be moved easily.

Self-filling troughs or buckets require a more complex set up.

If you'd rather purchase an automatic waterer there are plenty to be found on Amazon.

Whichever version you choose, make sure to keep your flock happy with fresh, clean water. You will be shocked how fast a flock of birds drinks a bucket of water. If in doubt, always leave more than you think you will need.

Water should be checked at least once a day. More often in extreme heat or cold. In the hot summer months, the turkeys will consume water faster. If you really want to spoil them, add a block of ice to cool them off.

In the winter, if you live in an area that experiences freezing temperatures, you will need to either fill their water multiple times a day or come up with a plan to heat their water. We use heated buckets in the winter. They come in various sizes, but all work the same. Keep the size of your turkeys in mind when purchasing a bucket. Keep in mind, most heated buckets require electricity. There are a few versions out there that are powered by small solar panels.

If you buy a large bucket while your turkeys are still little you must create a step or two for them to reach the top of the bucket and you must place a brick or stone inside, so they don't fall in and drown.

Bottom line. Without water, your turkeys will die.

Feed

What is feed? "Feed" is the food we give our farm animals. For turkeys there are four types of feed.

- Crumble: Crumble is basically a pellet that's been broken up.
- Pellets: A dense pellet or piece of food.
- Scratch: A mix of grain or seeds to create a "treat."
- Scraps: Any food you don't eat but turkeys find delicious. Fruit peels, vegetables, or stale pieces of bread make great treats for the turkeys.

Chicken feed or turkey feed?

Chicken feed and turkey feed are not the same. Chicken feed has a lower percentage of protein than turkey feed. Due to their quick growth rates, turkeys require a much higher amount of protein in their diet. They also require different amounts based on their age.

Poults require 30% protein crumble until they are eight weeks or older. Then they drop to 20% protein crumble the rest of their lives.

Once they are big enough to show interest in the pellets, you can feed them those as well as kitchen scraps.

What if my flock is mixed with turkeys and chickens? Can I feed them chicken food?
No. Not if you want healthy, full-sized turkeys. Turkeys require much more protein in their diet and chicken food just doesn't offer enough. While chicken feed or stock grain most likely will not be fatal to the turkey, it will not give them the proper nutrition developmentally.

Can I feed chicken food to my poults?
Layer or Breeder food for young turkeys is not wise. This food is too high in calcium for young poults or jakes and can cause all sorts of problems and/or death.

Feed Storage

All feed, except scraps, should be kept in a cool, dry location. Plastic bins with lids and metal trashcans work well. They keep the moisture and pests out.

Mice and rats love turkey feed. Wherever you have chickens and turkeys, the mice will soon follow. Farms have barn cats for a reason. Mice patrol!

Grit

Turkeys do not have teeth to cut up or grind their food. They swallow it into the gullet where stones and hard pieces of shell located in the gullet pouch break down the food before it's digested. "Grit" are those small stones and shell. Birds that are prone to foraging tend to pick up stones on their own. Turkeys in pens or clean areas do not have that luxury and must rely on their handlers to provide grit for them.

Treats

The list of treats for turkeys is endless. Like chickens, they love a good snack. Meal worms, fly grubs, and crickets are a favorite. Fruit is always popular too. Even pieces of meat or seafood scraps are popular. Nothing makes our flock come running like a glimpse of the scrap bucket. They see it from across the yard and come running!

What Foods to Avoid

There are many human foods to avoid due to toxins and substances that turkeys cannot tolerate or digest.

Do not feed turkeys the following human foods:

o   Avocado skin and pits. Avocado contains Persin, which is toxic to turkeys.

o   Citrus juice and skins. No oranges, lemon, or lime.

o   Don't give turkeys any edible containing salt, sugar, coffee, or liquor. No stimulants.

o   Uncooked or raw beans. Uncooked raw or dried beans contain hemagglutinin, which is poisonous to turkeys.

o   Raw potato skins. The rule is no green potato skins but best to play it safe and keep potatoes off the list completely. Raw green potato skins and eggplant contain solanine, which is poisonous to turkeys.

o   Onion. Large quantities of onions can be harmful to turkeys, affecting their red blood cells, causing hemolytic anemia, or Heinz anemia.

o   Peanuts. Old damp peanuts that may contain aflatoxins (mold).

o   Leaves of rhubarb, potato, or tomato plants.

The following can be fed in moderation:

o   Spinach. Spinach can interfere with calcium absorption.

o   Iceberg lettuce has little nutritional value and can cause diarrhea if too much is ingested.

o   Rice, pasta, and bread should be limited, especially white starches.

Other things to keep in mind while feeding turkeys are that they are curious by nature and will forage constantly.

Be mindful of mechanical parts left on the ground. Especially shiny pieces of metal which are attractive to birds but can be fatal if ingested.

When working on an enclosure or in the yard, be mindful of where you put screws and nails, or you might be short a few and end up with an emergency vet situation.

Cleaning the Coop

If you free range your birds, you won't have as big of a mess, but turkeys who roost indoors or are confined to an enclosure require assistance from their human handlers to clean their areas.

We practice the deep litter method and clean the coop twice a year. Spring and fall work well for our flock. Spring allows the coop to be emptied and dry out for the warm summer months. We use pine shavings inside the coop and only use about six inches in the spring versus double that amount in the winter. Pine shavings work very well for our coop.

Pine shavings – Pine wood shavings work well. There are two sizes available, large and small. Small breaks down too quickly and becomes very dusty. Too much dust can create breathing issues for your turkeys.

Cedar shavings – Cedar wood should be avoided in avian living spaces because it can cause respiratory issues. If you use wood shavings for bedding, make sure you are not buying cedar shavings.

Sand – Sand is great for a turkey run but not for a coop. It doesn't allow any insulation from the cold and can become muddy when wet.

Manure and Compost

Beautiful nutrient rich turkey manure!

Turkey manure is a valuable natural fertilizer containing droppings, bedding, nitrogen, phosphorus, and potassium. It is a wonderful medium for growing a garden or boosting the growth of plants and trees but should be used with caution and in the correct method.

How do you make manure from turkey droppings? It occurs naturally in their coop with time, or it can be helped along if you have a free ranging flock.

When droppings accumulate in the coop or turkey pen, they tend to pile up quickly. Especially if you are following the deep litter method of adding shavings to your coop instead of cleaning it out. Turkey droppings are four times the size of chickens' droppings and pile up quickly. When it's time to clean out the coop and start again, pile the mix of droppings and bedding into a compost bin or simply in a pile.

A composter can be turned which adds air into the mix. A compost pile needs to be turned or mixed up every few weeks. A pitch-fork works wonders for this!

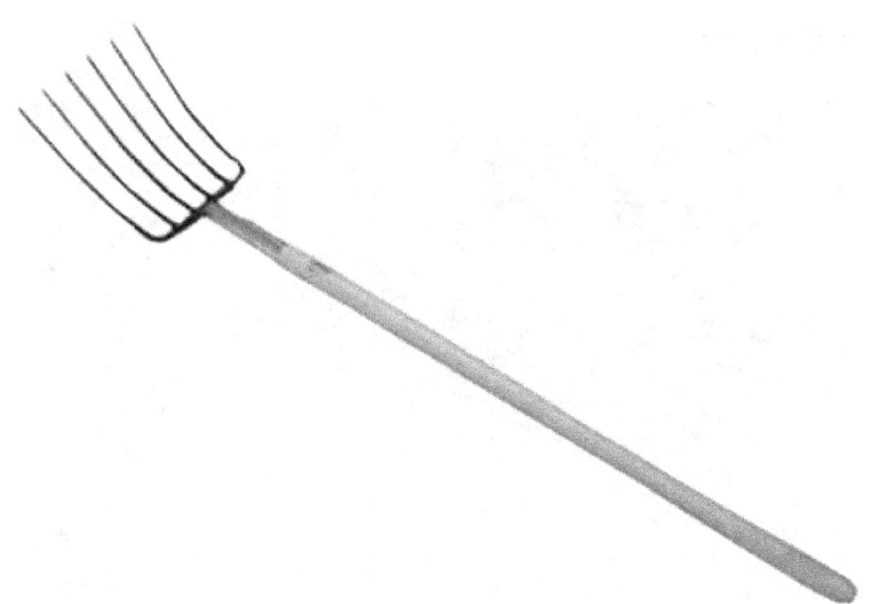

If your birds are free ranging and roosting somewhere other than a coop, you will need to locate their "piles" of droppings and collect them. A wheelbarrow is ideal as there will be a lot. Then you handle them the same way with a compost bin or compost pile.

Over time, the droppings and other scraps, grass, leaves, and other organic items put there tend to break down together. Water also helps this. In Colorado, we have a dry climate so it's necessary to add water as needed to help the decomposition occur. In states with high humidity, you don't have to add water and can simply "turn" the product.

In about three to six months, you will begin to see layers forming as all the items begin to break down. Congratulations you've made manure! Manure is the goal as it will be chock full of nutrients your plants will love. This is gold in the gardening world.

CAUTION

Do not dump turkey droppings on your plants thinking you can short cut the system and allow them to decompose on the plants. It doesn't work that way. Without time to break down, the turkey droppings are too "hot" with nutrients and will burn your plants. They will not grow correctly and may even die. Composting is the right way to go.

We could write a whole book on the benefits of composting but here are the finer points.

- Creating incredible nutrient rich soil that you know is clean of toxins or chemicals is a wonderful thing with too many benefits to count.
- Composting allows you to recycle scraps and yard waste that would otherwise go into a land fill or emit carcinogens into the air by burning.
- Composting turkey poop is part of a natural circle. The droppings go back into the soil to create more food and scraps. Naturally.

Turkey Eggs

Just when you think your turkey hens are never going to lay, they will surprise you with a beautiful egg. Rarely do you receive an egg each day, but you should receive a few every week. Turkey eggs are larger than chicken eggs. Double the size. They are very rich and are great to eat or bake with. Or allow them to sit where they are and watch them accumulate. If you have a male capable of fertilizing the eggs you will eventually have turkey poults.

Egg laying is mainly in March and April with peak hatching occurring in early May. Mating activities typically start in mid-March and nesting activity is high near the end of April. With the incubation period of twenty-eight days, most poults are present in the last week of May or early June. The exception to this is southern Florida, where the warmer climate gets the breeding process started around mid-February.

The egg laying season is in total contrast with chickens who lay continuously all year long with brief periods where they slow down but keep laying.

If you're wanting turkey eggs for a source of food, you may want to supplement your flock with chickens to ensure a steady stream of fresh eggs.

Cleaning Your Eggs

To clean or not to clean is a much-debated topic in the poultry world. Some like to collect and wash their eggs immediately while others debate that this rinses off the natural egg protectant that occurs when they are laid by the bird. Washing it off takes this protective coating away and potentially allows bacteria inside the egg.

But how does this happen in a solid eggshell? Eggs are porous. They have thousands of pores that open up to bacteria when washed. Personally, we do not wash our eggs until right before consumption but there are plenty of farmers out there who do.

Another option is to wipe them off with a soft cloth. If you've ever gathered a crusty, dirty egg, you know this isn't always the best solution.

Storing Your Eggs

How to store eggs is almost as hot of a topic as washing them. Do you store your eggs in a cold storage (refrigerator) or leave them at room temperature?

In the United States, grocery stores refrigerate their eggs. They are kept in cold storage for weeks before they even reach the grocery store shelf. Then, because we bought them cold, the consumers store them cold.

In Europe, eggs are sold at room temperature off the shelf. They are stored outside the refrigerator.

Our home only refrigerates eggs if they happen to be older than a week but that seldom occurs. We sell or give them away before they are that "old." Our eggs reside on the counter where I can keep a tally on how many we have. You'll be amazed at how quickly they accumulate.

Trimming Claws, Spurs, and Beaks

A turkey is just like a dog or a cat in that its nails, beak, or spurs may need to be trimmed every now and then when they get too long.

Trimming the Claws

If claws are left unchecked, they can grow into the bird's foot and become painful.

Trim with caution. Don't take too much!

Like cats and dogs, turkeys also have a sensitive area made of soft tissue in the center of their claws known as the quick. Cutting the quick by accident can be painful and bloody, so it's important to be very mindful about how much you trim.

You can make the process of nail trimming easier by soaking their feet in warm water or cleaning them thoroughly with a damp rag prior to trimming. Soaking softens the nails which makes them much easier to clip without the nail splitting. Clean toes also make the quick much easier to identify! Other ways to find the quick include looking underneath the nail rather than at the top or sides or using a flashlight right against the nail to illuminate the interior.

Hold the turkey in your lap, making sure to keep their wings secure. They should calm down a bit after some secure, gentle holding. If by yourself, use one hand to hold their foot and the other to do the trimming. Secure their toe with your thumb and forefinger to hold it still. They might be touchy—this isn't out of pain but because of distaste toward their feet being touched.[5]

---

[5] https://opensanctuary.org/article/how-to-trim-a-turkeys-nails-spurs-and-beak/

Using a cleaned pair of pet or human toenail clippers, trim a little bit of the turkey's nails at a time, not trimming past about one-eighth inch as you go to prevent accidental quick cutting. Each time you snip, look at the remaining nail. If it changes color, you're very close to the quick and should stop trimming. If the nail is still too long and you're near the quick (as it grows and recedes depending on overall nail length), you can wait a few weeks for the quick to retreat and then trim a little more, repeating the process of trimming and waiting until their nails reach a safe length.

In some instances, after you've trimmed the nails, you may need to file down the remaining rough edges with an emery board or nail file to protect the turkey from injuring themselves when they scratch. Make sure to clean nail clippers if they become dirty or contact blood from hitting the quick before using them to trim another animal's nails.

If you do accidentally draw blood, apply an astringent like a styptic pencil, styptic powder (such as Quick Stop), alum, or witch hazel. You can also dip the wound in cornstarch or flour to encourage natural clotting. Lacking these tools, you can also use a piece of toilet paper as if you'd nicked yourself shaving! If the bleeding doesn't stop, you can use the tip of your finger to apply pressure for up to a minute and repeat until any bleeding ceases. If you're going to return a turkey with a nicked quick to the flock, be sure to wait until the blood is completely stopped and cleaned to prevent exacerbation of the injury by other turkeys.

Trimming the Beak

If you need to trim or file a turkey's beak, how overlong it's grown will determine what technique to use. A small amount of overhang of the top beak is normal in most individuals. If the upper beak is just beginning to get too long, you can use a fingernail file or emery board to file it down a little bit. If it's gotten past that point, you can use cleaned pet or human nail clippers to very gradually trim down the excess in a similar fashion to their nails. Their beak also contains a quick that you need to carefully avoid snipping at risk of blood and pain. The part that needs to be trimmed is typically colored a little lighter and is more translucent than the rest of the beak. If you're concerned, look in the turkey's mouth to quickly discern where the beak's quick ends. Make sure to hold their head securely when trimming. Beak trimming is much easier with a helper holding the bird!

Be careful! Trimming the toenails to the quick is one thing but trimming a beak to the painful quick is another. Trimming too much can result in a lot of pain for the bird and make it extremely difficult to eat and function normally. If this occurs, you need to watch your turkey carefully. Use extreme caution when trimming beaks and go slow. Better to trim multiple small pieces to avoid trimming too much.

In most cases the upper beak will require trimming but there are rare cases of "scissor beak," a genetic deformity, or having been severely debeaked. You would treat both the upper and lower beaks the same way. With caution.

Trimming Spurs

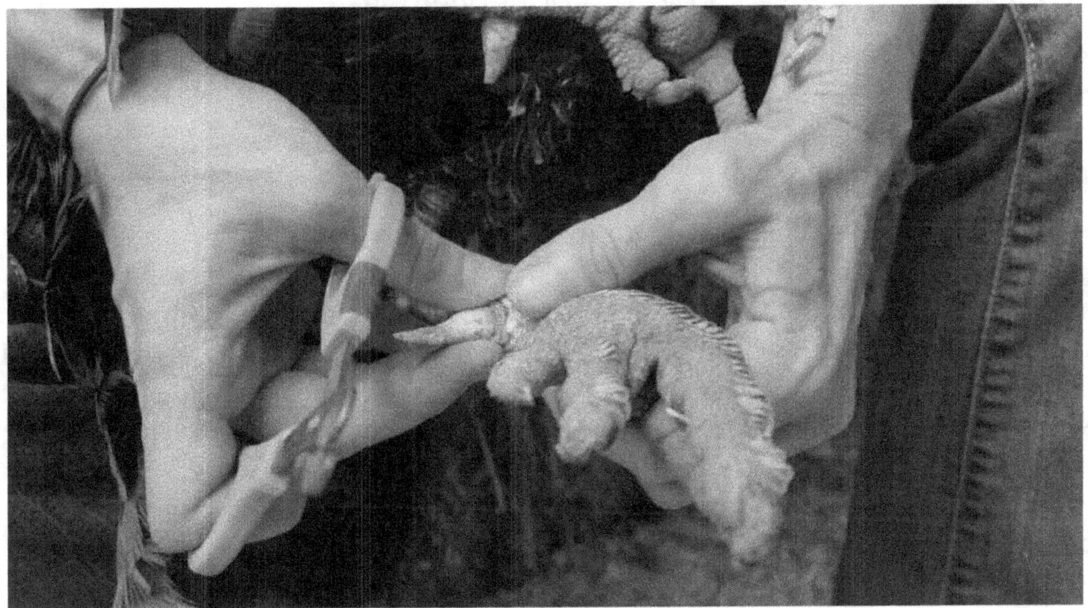
Be careful with the spur!

If you need to trim a turkey's spur, there are several options. You can blunt it with cleaned wire cutters, nail clippers (preferably pet), or even a Dremel sander with the proper sanding attachment (when done correctly, this method is the least likely to cause the spur sheath to crack). We personally like the Dremel method. It takes a little bit longer but seems to be cleaner without the risk of cracking a spur. Ouch!

Like their nails, the spur contains live tissue known as the quick. When trimmed too close to the quick, the bird will be in extreme pain. Avoid cutting the quick!

Hold the turkey's foot securely while trimming the spur to minimize risk of injury. It's also a good idea to stabilize the base of the spur and ensure you are not putting undue pressure on the spur which could cause damage or even cause the spur to break, which is both painful and bloody. Once trimmed, you can blunt the edge with a nail file or emery board.

How often will I need to trim claws, spurs, or beaks?
As often as needed. Keep an eye on your birds' feet and beaks. We're not saying you need to set up a turkey pedicure station every month. When it becomes noticeable, it's time.

Preparing Your Turkeys for Cold and Hot Conditions

Like any animal, turkeys need protection from the elements. Unlike other animals, sometimes you need to save them from themselves.

Turkeys in Cold Conditions

Albuquerque rarely gets below 20 degrees Fahrenheit at night. Almost never. Most of the animals on my brother's farm seemed to know when bad weather was coming. The chickens went inside to the safety of their roost. Even the goats stayed within the warmth of their shelter. Imagine my brother's surprise when he looked out and saw Mr. and Mrs. Belvedere with their heads tucked under their wings, freezing to death on the edge of the metal trailer. Out of sheer habit, stubbornness, or stupidity, the turkeys were out in the middle of a freezing storm. Wind and snow howled around them, and they were already covered with a layer of snow. Much to their surprise, he carried each one to the chicken coop and locked them in before they froze to death.

Wild turkeys roost in trees in any elements. Hot, cold, wind, and rain…they weather it all. There are no studies to show how many will die from exposure, but it would be interesting to know. Barnyard animals seem a bit more sheltered. Even though animals could survive it doesn't mean they should. Why not make their lives a little bit better if you can?

Sometimes you must step in and act for them. Turkeys behave on instinct. Please don't set them up for failure. Here are a few tips to help your flock remain healthy in the severe cold.

Best Turkey Breeds for Cold Climates

Some turkeys are better suited for cold climates. When picking out your flock it's so easy to get drawn into a certain color or size. Please be mindful of which breeds are better suited to your area. Birds that do well in the Deep South may not survive in Michigan or require special treatment to thrive.

Cold weather turkeys include:

- Standard Bronze Turkey
- Blue Slate Turkey
- Narragansett Turkey
- Bourbon Red Turkey
- White Holland Turkey

Notice a pattern? These are all large breeds. They have enormous bodies and copious amounts of feathers to help regulate their body temperature.

Turkeys are most comfortable in temperatures between 40–75°F. Turkeys are designed to cool themselves through their breath, unlike other animals that can sweat in the heat. Therefore, high heat temperatures pose a greater threat to them than cold temperatures.

How do you know if your turkeys are too cold? Look for signs of distress. Not eating, lethargic behavior, or not moving at all are all clear signs. Turkeys naturally like to be outside. If they are spending a lot of time in the coop or shelter, they are most likely too cold.

Coop Setup and Care in the Winter

Your turkeys do not need the Taj Mahal. They can withstand strong winds and temperature shifts as long as they have a few basic requirements.

Wherever they winter over, it's important to have dry bedding material. Pine shavings or straw are excellent at retaining heat and moisture. Stay away from sand as it has zero heat retention.

A roof over their heads and a dry draft free area are ideal.

Cleaning the Coop in the Winter

If your turkey is spending a lot of time "weathering in," you will need to help them by cleaning out their enclosure from time to time. How often depends on the number of birds and personal preference. Personally, we practice the deep litter method and add bedding to the top every few weeks. The layers of soiled bedding add up to make a great insulation barrier. The coop gets cleaned out in the spring and fall. Everything we take out gets put into the compost pile.

Nutrition in the Winter

As long as turkeys have food and water, they can survive just about anything. Their bodies require nutrition to fuel their metabolisms. This is a turkey's natural heat source. I always add a little extra corn to their diet in the winter. Corn is harder to digest and makes their bodies work a little harder which creates a little more heat.

Water is a necessity and should be monitored closely in winter. Just because there's a water source in their indoor enclosure doesn't mean it hasn't frozen over. Turkeys will not peck through the ice and will require your assistance in breaking the ice up.

There are all sorts of tips and tricks to keeping a bucket of water from freezing but a lot of them are unreliable. You either check their water multiple times a day or find a reliable source. We use heated water buckets. They require electricity and cost about $45 but they are by far the most reliable method of winter watering. They have an internal temperature gauge that switches the bucket on once the temperature drops to a certain point. Then it heats the water just enough so that it doesn't freeze. Our buckets have been in use for a few years. We use them for all our animals.

Caring for Your Turkeys in Hot Conditions

Remember, turkeys are most comfortable in temperatures between 40–75°F. Turkeys are designed to cool themselves through their breath, unlike other animals that can sweat in the heat. Therefore, high heat temperatures pose a greater threat to them than cold temperatures.

We can't say it enough. Heat can and will kill your birds. Adequate shelter and water are the key to the survival of your turkeys.

Water

Like most living creatures, turkeys require water to survive. Making sure they have a clean water source readily available is the number one thing you can do for your turkeys. When it's hot outside, they will consume more and rely on you to provide it. Buckets, a water trough, or a natural pond are all great sources. Just remember to fill them.

In extreme heat, we freeze water in gallon milk jugs and then place the frozen jugs in their water troughs. The turkeys and chickens seem to enjoy the cool water. Wouldn't you?

Shade

Turkeys like to roost at night. The same place they roost at night will most likely offer them shade during the day. This could be a tree, barn, loafing shed, large bushes, or in Mr. Belvedere's case, a large trailer. Even if they roost indoors, they will most likely come outside during the day. They will find the shade if there is some to be found. Help them out and build your enclosure under natural shade sources. If there aren't any, create some. Poles with military camo netting or shades can be put up. Plant some fast-growing trees or bushes. Or build a structure that offers protection and shade. I've seen some very creative coops. You're only limited by your imagination.

Alexandria, Louisiana, is home to Inglewood Plantation. They grow organic natural beef, pork, and chicken, among tons of vegetables and fruits. My favorite part of their farm is the chicken bus. They have a huge school bus that has been modified to house hundreds of egg-laying chickens. The bus gets moved every day or so to a new bug filled grassy spot on the farm and all its residents move with it. Then they place a net around the perimeter and allow the ladies to free range to their hearts' content. The bus is also high enough to create a natural cover protecting them from flying predators and the hot sun. Brilliant!

The Chicken Bus at Inglewood Farms

Molting

Grumpy turkey during a molt.

Molting is a part of life for turkeys. Turkeys will lose their feathers annually. This usually happens during the spring or fall seasons and is called a molt. During this period, your birds may lose a large portion of their feathers. Sometimes all of them! The bad news is they will look their absolute worst. The good news is their molt is only temporary. This natural process lasts between four and ten weeks as old feathers drop, and new feathers emerge. Then your birds will be handsome again.

If your turkey is losing feathers during a non-molting period, consult your veterinarian.

How to Care for Your Turkeys During a Molt

Keep caring for them like you would normally. Molting is a perfectly healthy process. There is no need to change their diet or provide extra nutrients, but you can offer additional protein treats if you'd like. It can't hurt.

Since they are exposing so much skin, it's a good idea to have a shade source as well.

Reduce Stress

Molting is generally not a good time to stress out your birds. Aren't they going through enough? Maybe now is not the time to switch enclosures or trim their beaks or claws. Give them a few weeks.

Do not pick them up or touch them right now. While their molt does not hurt them, it is painful to the birds when you handle them while they have new feathers growing in. It's best to leave them alone and allow nature to run its course.

Part 4: Troubleshooting

Common Health Problems

Diseases and Infections

Bumblefoot

Bumblefoot occurs when a turkey's foot develops a cut or steps on something which causes the area of the foot to become infected and very uncomfortable for the bird. The area hardens and forms a "kernel" which continues to grow until removed. If its not removed it can spread to the rest of the turkey and cause it to die. In most cases, it's easily removed by soaking the turkey foot in warm water. This softens the hard kernel enough to extract it with tweezers.

Botulism

Botulism in turkeys is just like humans. It comes from contaminated food. Turkeys who get it are typically fed food scraps that have rotted. Meat is a common carrier, but vegetables and grains can be contaminated as well. It is not contagious, but if one turkey got it, then you must question whether the entire flock was exposed to the same contaminated food source.

Symptoms: Botulism causes spasms and tremors. If you see your turkey suffering from a spasm or seizure, you must contact your vet immediately.

Your vet will have medication that can cure it and a plan for what to do if other turkeys show symptoms. If left untended, the bird will have trouble breathing, suffer paralysis, and die a painful death.

Thrush

Thrush is another food source illness. It is a fungal infection from contaminated food or water.

Symptoms: The turkey will want to eat more than usual and will be lethargic. They begin to have a crusty vent and a thick white liquid in their crop.

Tell your vet and ask for an anti-fungal medication. It is imperative that you clean the coop and all food and water sources. Make sure that you don't have a bag of food that has gone moldy so that the problem doesn't come around again.

Always keep your turkey feed in a clean, dry container, protected from water and humidity.

Fowl Cholera

This is probably not a threat so much in a suburban setting as it is a zoonotic bacterial infection from wild animals, but it's worth noting if you visit a zoo, wild animal habitat, or visit outdoor areas where wild animals or birds live, like the mountains. This is also a great reason to use one set of shoes for your farm chores and coop maintenance.

Symptoms: Diarrhea, swollen combs, and wattles (and possibly discoloration). There can be mucus coming from the nostrils and difficulty breathing. Some turkeys may look like they are having a hard time walking.

This is a fatal disease, and there is no cure. The bird needs to be put down as, once infected, they would still be a carrier even if they recovered. There is a vaccine for it; make sure that your turkeys have it by either injecting them yourselves or ordering them with it when purchased. If administering it yourself, it can be purchased from veterinarians or feed supply stores.

Parasites

Parasites that can multiply on your turkeys include mites, lice, worms, fleas, and ticks. We are going to look at the most common ones, discuss prevention, early warning signs, and the required treatment.

Red Mites (Dermanyssus gallinae)

Red mites are almost invisible to the naked eye, and they have a short lifespan of only about seven days, but they multiply in the tens of thousands. They are most active from May–October and go dormant in the winter, but they **do** live through winters inside the coop for months, so don't expect cold weather to solve the problem.

Red mites can be passengers of wild birds, but in a backyard coop, the most likely way they will arrive is through a new bird or from clothing or a pet (e.g., a dog) that has been in a place where they have set in. You could, for example, visit a local breeder or animal swap and bring your dog with you, tour the farm, bring mites on your clothing, and bring them home to your turkeys through both your clothing and your dog. Your turkeys will eventually be covered with them all over, particularly around the vent.

Symptoms: Red mites feed on your turkey's blood. Your turkey will become anemic. Their wattles will fade. As the infestation increases the hens will stop laying eggs. Sometimes they start refusing to roost at night because the red mites hide in the cracks of the coop during the day and jumps on the birds at night.

Red mites are probably the most common problem of poultry raisers, and prevention is a lot easier than treatment.

To prevent red mites, you will need to empty, scrub, and clean your coop weekly. When you scrub it, there are two suggestions that make it easier to really get the mites out of the coop:
- Use a steamer that is used for stripping wallpaper when you clean your coop.
- Use a hose with a nozzle and as much jet pressure as possible. Some chicken and turkey raisers even choose to invest in a pressure washer so that they can really get into those cracks and crevices.

If you use DE, rub it into the cracks of the coop and on the perches. Using DE as a dust bath as well as in coop cleaning has been successful for me so far; it is all I do. If you decide not to use DE, another suggestion for killing red mites in the coop is to rub a mix of paraffin and Vaseline in the cracks and corners of the coop. This suffocates them.

If you do have an infestation, then besides getting rid of the mites, consult your vet about vitamin supplements in the turkeys' water to help them recover fully and more quickly.

This is the best information I have found about red mites. It helps you understand the life cycle of the mite and how best to clean and when for maximum effect.

Scaly Leg Mite

Scaly leg mite is the second most common mite. They are easier to deal with but may take time for the turkey to fully heal. 0

Symptoms: Since they are so small, it will probably take several weeks before you'll notice that there is an infestation. They live in between the scales of the turkey's legs. What you finally see is a build-up of the mite poop, fluid oozing from the legs of the turkey, and their expressions of discomfort. The mites spread from other birds, so it is likely that all the turkeys will have it if one does.

A common and simple treatment is to apply Vaseline to the legs, rubbing it into the crevices made by the scales. Mites need air, so the Vaseline will smother and kill them.

Don't forget to throw away all the bedding and to clean the coop just like you would for red mites. They don't live in the crevices of the coop in the same way as the red mites do, but they may be lurking, so you must do a big clean when you treat your turkeys.

Mites of the Feather Shafts and Bases

The red mite and scaly leg mite are the two most common mites. Other mites are:
- **The Depluming Mite** – this one burrows into the feather shafts; the bird's skin oozes as though it had a wound, and they may begin to pluck their feathers to try to alleviate their discomfort.
- **The Northern Fowl Mite** – also found at the base of the feathers, but instead of burrowing into the shaft, they are around the base of the feather.

If you see signs of either of these two mites, you must contact your vet for immediate treatment recommendations. The key to success is early detection. The checking of the skin around the feather shafts should be your regular weekly check-ups. Furthermore, an example of why it is so important to handle and take care of your birds at a young age is to ensure that they are tame and will not be frightened by your examinations.

Fleas

Some locations are more prone to fleas than others. Fleas can act like the northern fowl mite; they will be found at the base of the feathers as well as under the wings and around the vent. They can be seen by the naked eye, and like your dog or cat, your turkeys will let you know that they are uncomfortable. Talk to your vet about the diagnosis of which flea it might be and follow their recommended treatment.

Ticks

Ticks will attach to your turkeys as happily as they will do to you or your pets. If you live in a place where ticks are likely, then you must inspect the skin of your birds for any attached ticks.

Symptoms: If a tick is attached to a turkey, the symptoms of the bird being the host of the tick will indicate that the crown and the wattles will be pale, as well as low energy and fewer eggs. These symptoms could be related to several issues, but if it is tick season, include a careful examination of the skin in your routine. If you find one attached, you can use tick tweezers as you would for yourself or another animal.

If you have a hen with one or more ticks, do a big coop clean, throw out the bedding and clean all corners, crevices, nooks, and crannies.

Worms

There are a lot of different kinds of worms, and it takes a lab test to determine which one it is. Instead of living outside the body like mites, fleas, and ticks, worms live in the gut. The worm eggs are in the poop of the birds, so as the flock pecks around, there is a circular pattern to the infestation: Worm egg hatches in the gut, lays eggs, eggs are pooped out, birds peck in poop, get more eggs, and that's how the cycle continues.

Contact your vet for a worm test if:
- Your birds are short of breath, especially stretching their necks and gasping for breath.
- The yolks of the eggs are pale.
- Turkeys are eating a lot but losing weight.
- They have diarrhea or loose stools.
- You notice signs of lethargy.
- The birds have faded combs or wattles.

As with most problems, prevention is the best cure. Worms are a fact of life in the spring and summer but ensuring that you do not have an ideal worm hatchery environment can do a lot better for you and your turkeys. Keep both the coop and the run clean and dry. UV light kills worms, so rotating the ground or dirt will help a lot on a sunny day. Change the bedding often and keep it clean and dry as well.

Personally, I would want to get a lab test and confirm if that which I'm dealing with is a worm. After that, I would use whatever treatment the vet recommended. Some turkey raisers start with putting some apple cider vinegar or crushed garlic in the water to see if the problem can be alleviated.

Some poultry farmers swear by pumpkin seeds as well. Toss a pumpkin into the run and watch the flock devour it. The seeds clean out their intestinal tracks and the birds get to have some fun.

If you do have a bird with worms, it is good to treat your turkeys and then move them, so they aren't pecking in the same place as they were. In the meantime, you can expose that area to the sun by turning the substrate and keeping it dry.

Wounds

Wounds can be caused by pecking or fighting, by the talons of an aerial predator such as a hawk, an odd mishap around the yard, or toms being too rough on your hens when mating (a common occurrence).

Regular checks on your flock will reveal any wounds. Especially on white turkeys. The darker varieties will make it harder to detect.

Stop the Bleeding and Assess the Wound

Stop the bleeding with direct pressure as you would on anyone. When you have controlled the bleeding, you can assess whether the wound is deep or surface. A surface wound can bleed a lot, so have your gauze ready, keep the pressure on, and see what the problem really is. If it is a deep puncture wound or something serious like a severed wing or leg, continue to control the bleeding and call your vet. If it is a surface wound, you can continue to treat the wound.

Treat the Wound

It is good to have a first aid kit around for your flock just like you might for your family.

- Nitrile gloves for handling a bleeding bird
- Gauze
- Vetericyn Plus Poultry Spray
  - This is a non-toxic, anti-microbial spray that is safe to use on your turkeys Some recommend hydrogen peroxide, but since first aid certification classes teach you that peroxide is not to be applied to wounds and can be counterproductive to healing, we strongly recommend Vetericyn instead. (If the wound is around the eye, use an eyedropper or other applicator so that the spray doesn't go in the eyes.)
- Vetrap bandages
- A large clean towel dedicated to turkey wrapping when necessary

Isolate the Turkey While the Wound Heals

Your emergency backup turkey coop comes in handy again. Isolate your bird and apply the Vetericyn Plus three times per day. Care for it as you would a wounded family member or pet. Make sure the wound is completely healed before reintroducing the turkey to the flock, and make sure you watch the behavior of the other birds closely to ensure that she is safe to be mixed back in.

I find it handy to keep the backup cage near the existing enclosure. It needs to be close enough the flock can see the injured bird, but not so close they can peck or trample it. As long as the flock can see the injured bird you shouldn't have to reintroduce the bird back to the flock.

Sour Crop

As we discussed in the section about feeding grit, the crop in a turkey is a pouch in the esophagus where food is stored and broken down before it moves to the stomach. The grit provided to the turkey helps break down the food since they don't have any teeth. Along with the grit, there is also a healthy amount of bacterial activity that makes the food digestible.

Sour crop occurs when the bacterial activity goes out of balance, and there is a yeast infection. The turkey's breath will be horrible (hence the name), and the food will not move to the intestines; it will be stuck.

There are directions available for applying massage to the turkey to release the fluid, but this is an activity for only an advanced turkey raiser. Even then, it has a high risk if not performed correctly.

If you think your bird has sour crop, take it to the vet and follow their treatment and advice. At the very least, watch a YouTube video and attempt to learn how to treat the impacted bird.

Egg Bound Hens

A hen that is egg bound has an egg trapped within her that cannot be laid. This needs urgent attention and can lead to death within twenty-four hours.

Symptoms: You may see the hen straining with no egg coming out or going in and out of the nesting box without laying an egg. She will not be able to poop. She may waddle uncomfortably and stop eating or drinking. When you pick her up, she is likely to have a hard abdomen.

Sometimes you can deal with this yourself by placing your hen in a warm bath for one-half hour to relax the muscles and allow her to lay the egg. Gently applying some lubricant to the area after the bath can help. Having some nitrile gloves around for this purpose is useful.

If the bath doesn't work, get her to the vet. It is not recommended that you try to release the egg yourself unless you are a trained vet technician or have the training and experience to do this without harming or killing the hen.

Broodiness

A hen is "broody" when her hormones kick in to give her the maternal instinct of sitting on eggs to hatch. This is the same instinct that makes her a good mother. It is *not a desired trait* for turkey raisers who do not want to hatch and raise their own chicks and just want eggs.

Broody turkeys are very intimidating as they puff their feathers, hiss, and gobble at you the minute you step toward them and their "nest."

In the section regarding hatching chicks, we covered caring for a hen who has fertilized eggs that you intend to hatch. This next section will deal with broodiness as a problem when a hen is sitting on unfertilized eggs.

How to Tell if a Hen Has Gone Broody

A hen that has gone broody sits on her eggs in her nesting box or nest, fluffs her feathers out, and will not move except once a day or so to poop and get some food and water. She will become much less tolerant of visitors, even hostile and aggressive. She is likely to hiss, peck, and even bite when you try to collect her eggs. If you take her out of her nest box, she will run back obsessively.

The temperature of a broody hen goes up so they can keep their eggs warm. If left, she will start to pull out her chest feathers (they do this to line the nest and heat the eggs directly).

How to Prevent or Discourage Broodiness

You may not be able to completely prevent yourself from ever having a broody hen. There are ways to discourage it, though, making it a less likely occurrence.

Choose Wisely

The first prevention is to choose breeds who are not known for going broody. Heritage turkeys tend to go broody more often than other breeds, with good sitting tendencies and strong mothering instincts. The breed least likely to be successful is the Broad Breasted varieties. The broodiness has been commercially bred out of them.

Change of Scenery

The second discouragement of broodiness is to remove the encouraging environment where the hormonal process and maternal instincts thrive. That would be a nest with eggs. Understand the cycles and seasons of the turkeys you have. Turkey hens tend to lay more in the spring. Broodiness is most often an issue in the spring and summer. Some hens will lay during the winter, so they need to be watched for broodiness all year, but even these breeds will *tend* to go broody more during spring and summer.

What to Do if You Have a Broody Hen

There are many things you can do if you find out that your friendly, tame turkey is now a feathered demon who doesn't want to leave her nesting box and is suddenly hissing and pecking at you when you collect the eggs.

- Ask yourself if you'd like to hatch some chicks. If you have a breed of turkey that lays and a tom capable of fertilizing the eggs, now is your chance to hatch some chicks without having the hassle of an incubator!

If you don't want to take advantage of the moment and hatch chicks, then try the following methods.

A note on handling broody hens: Wear gloves! Some people wear leather gardening gloves, but I found that I needed to use my thick welding gloves that go up to my elbows. For some hens, safety goggles might not be a bad idea. At the very least, wear gloves when you are collecting the eggs as you are likely to get pecked. Turkeys are tall and strong. They are also very fast when they want to be. Better to be protected than not.

- Collect the eggs as soon as she lays them. Preferably immediately after she lays them.

- Block off her nesting box for the rest of the day so she can't get back in once you've removed her.

- Let her be broody. It takes 28 days for turkey eggs to hatch and the brooding cycle to complete. Another hormonal cycle kicks in to raise the poults, and if they are not there, then the hen goes back to normal. Some turkey raisers cleverly replace the real eggs with wooden eggs and just let the hen sit until she is done.

  There are four important things to note if you decide to do this:
  1) While a hen is broody, she is obsessed with those eggs and neglects her own health. She won't bathe or get exercise or even take in enough food and water if not encouraged. If you decide to let her be broody, keep her water and food within close proximity, and make sure she is forced out a couple of times a day to eat and drink.
  2) You may need to put her in a separate box as she may get aggressive with the other hens in the nesting box. Be ready with a "Plan B" for her nesting space.
  3) She's not the only one who can get aggressive. Broody hens are vulnerable to being bullied. When she does make her one trip a day out to get some food and water, she may be kept from accessing them and may even be pecked and wounded—another reason to separate her.
  4) Broodiness is "contagious." When one hen goes broody, others will likely follow. Letting her go through the cycle may mean that you have a henhouse full of broody turkeys going through their four-week cycles for weeks to months! Separating the flock would help with this and the other issues listed above.

- If necessary, provide a single wire cage enclosure for her until she has stopped being broody. The reason this works is that the wire on the bottom of the cage (that is off the ground) makes it uncomfortable, and she does not feel like there is a good place to nest. The air circulation will also cool down her chest and abdomen; therefore, her hormones will then shift, and she'll come out of her brood. This usually takes three to four days. You can check after three days to see if she's done. If she doesn't run back to her nesting box, then you know that the broodiness cycle has been broken. Make sure she has plenty of clean, fresh water and food.

- If you think you want to use this method, plan ahead and have a cage ready for her isolation.

Soft-Shelled Eggs

Soft-shelled eggs happen for a variety of reasons.

Young Hen – This is the most common reason soft-shelled eggs occur. The issue usually rights itself within a week or two. These eggs are found at the beginning of a young hen's laying cycle. It's not uncommon to find a soft-shelled egg or two before the hard eggs appear.

Nutritional Deficiency – Soft shelled eggs can also occur because of a nutritional deficiency. Soft shelled eggs are surprising when they happen but usually only a fleeting issue. While surprising, they normally aren't a cause for alarm but a sign that your flock isn't getting enough calcium and protein. Egg shells will only be thick and strong if your turkey has enough calcium and protein to create them that way. Boosting calcium and protein is an easy and manageable problem. Don't fret!

Quantity of Eggs – A turkey will lay a clutch of eggs (13-16 eggs). When the number of eggs increases, the turkey's body may not be able to keep up with the nutritional demand.

Old Age – Older hens may not be able to keep up with the nutritional demands of laying.

Mineral Absorption – Several diseases can inhibit the turkey's body from absorbing the minerals needed to create hard-shelled eggs. If your young, healthy, and otherwise fit turkey suddenly starts laying soft-shelled eggs, it may be a good idea to have them checked out by your local vet.

Stress – Like any animal, stress will cause a turkey's body to react. Moving, animal attacks, mates dying, or changing their roost placement can all cause stress with a possible side effect of soft-shelled eggs.

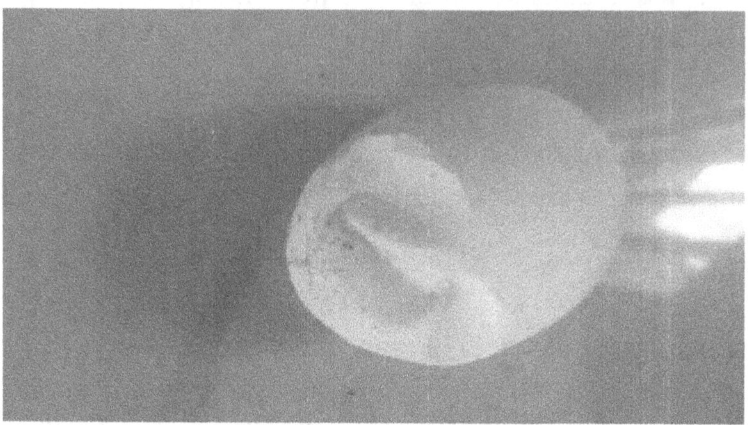

There are a variety of solutions to boost your turkeys egg production.

- Boost the turkeys' intake of protein and calcium.

- Oyster shell is an easy solution. It comes in bags ground up into small pieces and all you must do is place a bowl of it in the turkey enclosure or sprinkle it on their food. Personally, we find that our flock of chickens and turkeys ignore the bowl of oyster shell completely but happily gobble it up when mixed with their food.

- Supply a protein rich food. Earlier we discussed how turkeys require a higher protein percentage than chickens or other fowl. A mature laying hen requires 19% protein feed. If she's broody, she may not require that much, but her feathers will grow in faster with an additional protein boost.

- Supplement their normal food with protein rich treats like meal worms or grubs. They will love you for this too!

- Eliminate stress. Don't move their roosting spots if you can help it. Turkeys are social creatures. A flock of turkeys are happy turkeys. Happy turkeys lay better.
- Age is impossible to remedy but your hens can be helped the most by a calcium and protein rich diet.

The Pecking Order

A turkey flock is social by nature and creates a social ranking within the flock. This is called the pecking order. Aggressive behavior, age, and size determine the pecking order which is usually maintained until something happens to a bird and the order shifts. This can be caused by a bird dying, a battle where a lower ranking turkey defeats a higher-ranking bird, or a ranking turkey becomes weak and loses its position.

When a flock breaks up so does the pecking order.

Watch your flock and you will see set behaviors belonging to different groups. The gobblers tend to stick to themselves and can be found challenging each other for the top spot. Gobblers will walk in front of another gobbler and wait. This is a challenge. If the challenged turkey does not want to fight, he will turn around or even sit on the ground to avoid battle. Usually the flock will continue "as is." Sometimes that's not enough for the aggressor and he'll peck the turkey's head until he fights back or runs; a win for the aggressor.

Hens tend to stick with the hens and jakes flock with jakes once they leave the mother hens.

The pecking order determines a variety of behaviors. The birds on top tend to eat first, eat the best scraps, and grab the best roosting spot, which is usually higher than the other birds. Gobblers and hens who hold higher ranks tend to also be the watch dogs of the flock. They sound the alarm at danger and watch over the flock.

Bullying and Cannibalism

Flock bullying and cannibalism can occur. Turkeys seem to be more mild-mannered than chickens, but it still exists.

If a turkey suddenly appears lame or has an open wound, the other birds will begin to pick at the bird and wound. If the bird dies, its not uncommon for the other birds to cannibalize it until removed. If a member of your flock has an open wound, it's best to separate the bird until healed. Then slowly reintegrate it back into the flock. Most flocks do not take well to strangers joining the flock.

We have an enclosure that allows us to put up a temporary fence across part of the pen. It allows animals to see one another without being able to attack one another. They are used to seeing each other and when its time to take the wall down they are already used to one another.

What Causes Bullying

- Illness– Any sign of weakness seems to trigger the bullying.
- Broodiness– These ladies are grumpy and ready to rumble.
- Over Crowding– Too many birds in one area creates hostility and bullying.
- Stress– Stressed birds can and will bully each other.
- Boredom– Bored birds look for ways to stimulate themselves and if they have nothing to keep them occupied, they will make their own fun and harass their fellow feathered friends.
- Wounds– An open wound on a turkey is like hanging a sign that says, "Pick here!" The color of the blood is different which draws unfortunate interest.

What to Do About the Bullies

Flock bullies can be a big problem. Especially if they are stressing the flock or causing harm. Sometimes separating the bully for a day or two is enough to knock them off the top of the pecking order. This can be done repeatedly until another bird takes its place.

When a short separation isn't enough, permanent separation may be in order.

Some people try hosing the bird, throwing things at it, or chasing it off. These do not work. They only create hostile turkeys.

Natural Predators

Mountain lions caught on a trail cam in 2020.

There are all kinds of predators or animals who would like nothing better than a nice turkey for dinner. There is nothing worse than discovering something made of a meal of one of your birds. Not only is all your hard work raising and taking care of them for nothing, but you feel guilty. You're their protector and you failed. This is why creating a predator proof enclosure is so important. In a free-range flock, the flock depends on their own senses to protect them, and it works to a point. If you are planning on allowing your flock to free range and roost you might want to plan for some of your birds to die or disappear. It happens.

Natural predators to turkeys will depend largely on where you live but include the following animals.

Dog

Cat

Fox

Mongoose

Raccoon

Bear

Coyote

Cougar

Bobcat

Possum

Eagle

Hawks

Owl

Some of these creatures can be thwarted by a strong sturdy fence. Others require an electric fence or deep fencing to avoid digging under. It's amazing what animals will do for a free meal.

Dogs and cats are listed because even though they are domesticated, they still pose a threat. While turkeys are young and small, they are easily killed by cats. Dogs always pose a threat. Your dogs may leave your animals alone, but the neighbor's dog may not. Once a dog kills a bird, they will come back to do it again. Entire flocks can be wiped out.

Use caution and planning to avoid catastrophe. The saying, "Good fences make good neighbors," is true when it comes to your flock.

Part Five: Processing Your Turkeys

WARNING!

··

THIS SECTION IS ALL ABOUT PROCESSING TURKEYS FOR CONSUMPTION. IT CONTAINS GRAPHIC DESCRIPTIONS AND PHOTOS.

Killing a Turkey

Most people raise turkeys for the goal of eating them. If this is the case, you need to know what the best killing methods are. While these methods seem brutal, they are the most common, the quickest, and often most humane.

METHODS

Decapitation

This is the quickest and most thorough way of ending the life of a turkey. It is also by far the most shocking to someone who has never done it before. Due to the size and strength of a turkey, it's recommended you have help in holding the bird. Once it's contained, hold its head flat on a strong surface. A sturdy board or stump is perfect. An axe is ideal because of the weight and force needed to sever the neck. Once you commit to this you must follow through or risk injury to yourself or the bird. An injury only prolongs the animals suffering and that's not what you signed up for.

With strength and accuracy, you must swiftly chop through the neck. Once it is done, the bird's wings and feet will automatically begin moving. The wings will flap and it's a mess. If you've never killed a turkey or chicken before, this will be shocking to you.

Killing Cone

There are killing cones created for the sole mission of retaining a bird while this process is happening. You place the bird upside down in the cone and it's subdued and restrained. Then its neck is sliced, and it bleeds out. The cone keeps the wings from flapping and it's a little bit cleaner. A little bit.

Here's an in-depth list of what you will need to use the killing cone method.

You'll want to have all your supplies gathered and your slaughter area set up before you begin. A turkey-sized killing cone should be mounted to the side of a building, or a framework built out of wood. Here's what you'll need:

- **Knives.** Be sure your knives are sharp. Four to six inches in length is ideal. Two or more knives are needed.
- **Killing cone.** Find a turkey-sized killing cone at a farm supply store or order it from your hatchery. Some people DIY it with an orange traffic cone.

- **Buckets and pails.** A bucket or large plastic garbage pail underneath the killing cone and scalding area catches feathers and blood.
- **Water.** A hose with fresh, clean water is a necessity.
- **Table.** You need some kind of surface to process the birds on.
- **Scalding tank.** A very large pot or tank on a burner that can heat to 140 degrees Fahrenheit. Make sure the scalding tank is large enough to dunk and swirl the birds to remove the feathers.
- **Cooler or tank with ice.** A large cooler or tank filled with ice and cold water is necessary for chilling the birds after processing.
- **Paper towels.**
- **Cutting board.**
- **Plastic bags for storage.** There are heat-shrink bags available from farm supply stores that shrink tightly around the bird and prevent freezer burn.[6]

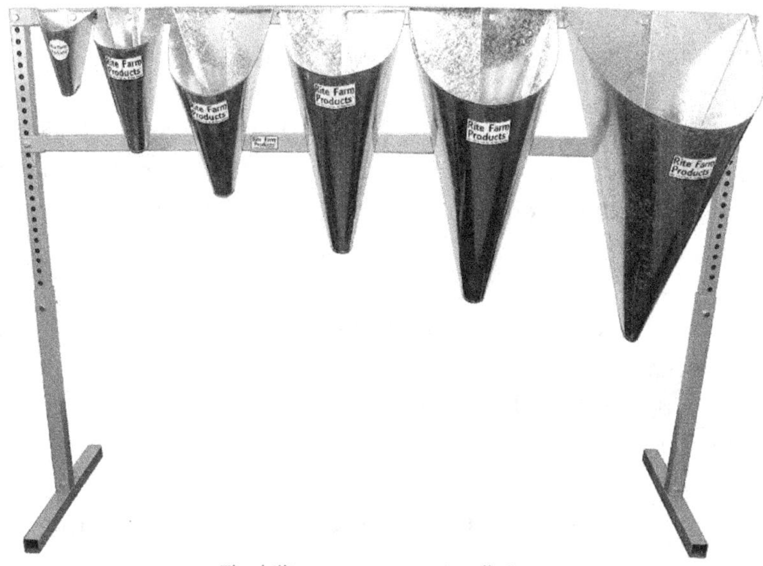

The killing cones come in all sizes.

PLUCKING

Once the turkey is dead you must pluck it. Removing the feathers is a tedious and necessary job, otherwise you must skin it and lose all that delicious skin.

---

[6] https://www.thespruce.com/how-to-slaughter-and-process-turkeys-3016807

To pluck a turkey, you will need a pot large enough to submerge the turkey in boiling water. Fill the pot with water and bring it to a boil. Trust us when we say this all needs to be done outside. It will be a mess.

Once the water is boiling, dunk the turkey head-first into the water. The goal is to loosen the pin feathers and not cook the bird. Dunk it for a few seconds and pull it out. All birds are different. When you can easily pull large feathers out, it's ready to be plucked.

Start in one area of the bird and start pulling feathers. You will be amazed how many feathers they have.

Some people save the feathers for decoration or to sell. There is a market for feathers. If you choose not to go that route, you can simply dispose of them.

Dark colored birds will have dark pin feathers. It's sometimes hard to get them all out. White turkeys have white pin feathers and are harder to see. Time and patience are required to pluck a turkey.

Once all the feathers are out its time to dress your bird.

Dressing a Turkey

Dressing a turkey is the fancy way of saying "gutting" or removing the innards. It must be done if you plan on cooking the bird whole or want the skin. If skin isn't important to you, skip the plucking and skin your bird, but you will still have to remove its innards.

Gutting a Turkey

Start by laying the turkey on its stomach on the board. Proceed to cut from the back of the turkey's neck to separate the esophagus and trachea. This process should free the two body parts so that you can have an easy time handling it. If the turkey does not have a neck, you can always skip this part.

You can now flip the turkey to lay on its back so that you can make the cuts that will make it easy for removing the innards of the turkey. Start by cutting a horizontal slit between the breastbone and anus. The slit is supposed to be large enough so that you can easily insert your hands and pull the innards out. When making the slit cut, you must be careful not to puncture the innards.

Proceed to insert your hand into the slit to pull the organs out. You can now easily pull out the organs such as the heart, lungs, intestines, liver, and gizzard. Once you get them pulled out, you should properly dispose of them.

One quick tip for removing the intestines is that you must cut a circle around the anus as a way of loosening the digestive organs.

With the organs removed, you can set them aside as not all of them would be cooked.

Cleaning the turkey

Now that you are done with the gutting, you can rinse the turkey in the sink. You need to use high pressure for the water. The water under high pressure is great for cleaning the turkey inside where the organs have been removed. You must make sure that all the blood has been cleaned from the turkey before you can prepare it for cooking. You need to use cool water so that you minimize bacteria growth.

How and when you cut up your turkey is personal choice. I like to leave the feet on until it's gutted so there is something to hang on to. Wet turkeys are slick turkeys!

Once gutted, chop the feet, and trim the neck to make it look like one those pretty, store-bought birds. Once rinsed, it's ready to be eaten. Although, many turkey processors say the meat is better if you allow your bird to sit or "hang" for at least three days.

Preparing Your Turkey for Dinner

Roasting a turkey does not have to be difficult but for some reason it really makes people nervous. It may have something to do with turkey usually being the highlight of just about every Thanksgiving meal served in America.

The truth is, it's really not that hard.

It also yields a gorgeous, super juicy, perfectly cooked turkey.

The key to a perfect turkey is not to overcook it—that's what dries out the meat, and you want to carve into a juicy, moist turkey on Thanksgiving! So, just plan ahead to get the timing right.

Not only is this roasted turkey recipe fool-proof and easy to make, it will also exceed expectations as the centerpiece of your Thanksgiving feast!

The general rule is to estimate one and one-half pounds of turkey per person. (So, if you're feeding six people, buy at least a ten-pound turkey.) That way you'll have leftovers. Leftover turkey is amazing!

Do not attempt to cook a frozen turkey! You must thaw it out ahead of time. Allow one day in the fridge for every five pounds of turkey. A 15-pound turkey would need to thaw in the fridge for three days. Be sure to place the turkey on a cookie sheet or pan to catch any liquid the turkey may drip as it defrosts in the fridge.

Rinse the turkey inside and out. Dry it off with some paper towels, and then prepare it for the oven.

You can fill the cavity of the turkey with things that will give it flavor like salt and pepper and any mixture of onion, apple, carrots, celery, or citrus. The flavor will permeate the bird from the inside out.

Another method is to create a bread-based stuffing and fill the cavity and neck. The cooking time must be extended but it creates a delicious bird inside and out.

The only way to really tell if the turkey is cooked (165 degrees Fahrenheit), is by using a thermometer. Test the turkey right from the oven—if it reaches 165 degrees Fahrenheit, your bird is safe to eat.

After removing the turkey from the oven, let it rest for at least 15 minutes.

If you have a big metal roasting pan, great! If you don't, no worries. Just use a casserole dish big enough to fit the bird or buy a disposable foil pan (then you can throw it away after). Place a bunch of chopped veggies on the bottom of your pan. The veggies will act like the wire rack in a regular roasting pan by elevating the turkey slightly. Place the turkey right on top of the chopped veggies.

After the turkey is finished cooking there will be juice and browned cooked bits at the bottom of your roasting pan. You can reserve all of it for making turkey gravy. This produces the most flavorful gravy!

How to Prepare a Turkey for Roasting

First, remove the thawed turkey from the pan it was thawing on. Rinse the bird inside and out. Pat the turkey dry with paper towels.

Next, season the cavity of the turkey with salt and pepper. Stuff it with the quartered lemon, onion, apple, and herbs.

Tuck the wings of the turkey underneath the turkey and set the turkey on a roasting rack inside a roasting pan (or on top of a bed of chopped veggies—carrots, onion, and celery work well—in a disposable roasting pan). Tucking the wings prevents them from burning and helps the turkey sit flatter.

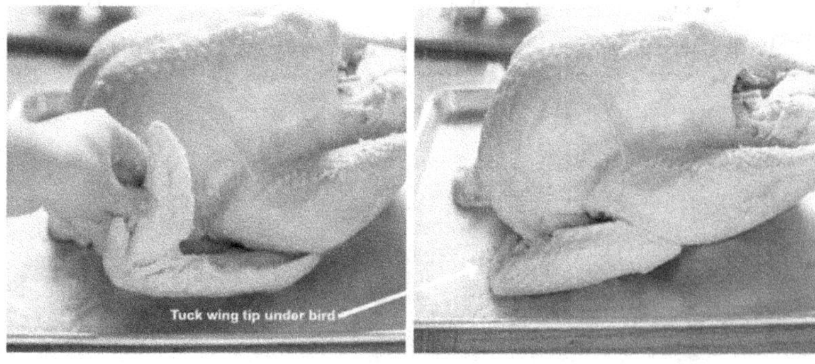

Use your fingers to loosen and lift the skin above the breasts (on the top of the turkey) and smooth a few tablespoons of the herb butter underneath. Use some twine to tie the turkey legs together. Then slather the outside of the turkey in the rest of the herb butter. That's it! You are ready to let your oven do the rest of the work!

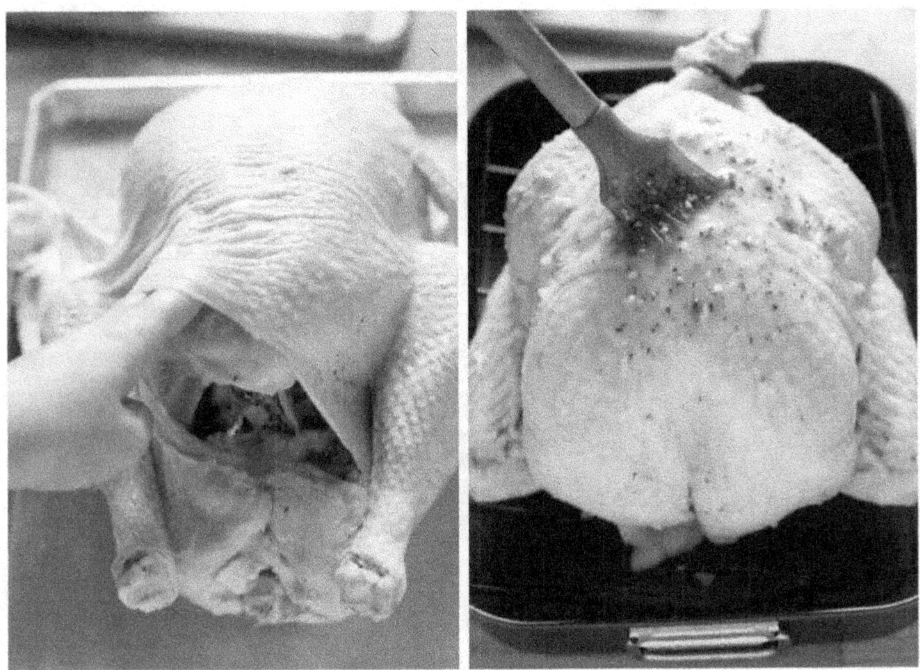

Remove the top rack of your oven and move the second rack down low enough for the turkey to fit. Preheat your oven to 350 degrees Fahrenheit and place the turkey in the oven.

The general rule for cooking time is 15 minutes per pound for unstuffed turkeys and 20 minutes per pound for stuffed. A 20-pound bird (unstuffed) would take five hours to cook. The same bird stuffed would need to cook for six and one-half hours.

Check the turkey about half-way through cooking. Use your oven light to see if the skin is golden, and then place a large piece of tinfoil over the breast meat of the turkey to help keep it from overcooking.

A great way to tell if your turkey is done is if the legs and wings pull away from the body easily and the juices run clear. If the turkey juices are pink or red, it needs to cook longer. Just keep an eye on it.